FUNDAMENTALS OF ENGINEERING GRAPHICS

Joseph B. Dent, P.E.

W. George Devens, P.E.

Frank F. Marvin

Harold F. Trent, P.E.

Division of Engineering Fundamentals
College of Engineering
Virginia Polytechnic Institute and State University
Blacksburg, Virginia

Macmillan Publishing Company
New York
Collier Macmillan Publishers
London

Macmillan Publishing Company
866 Third Avenue, New York, New York 10022

Collier-Macmillan Canada, Inc.

ISBN 0-02-328490-0

Printing: 2 3 4 5 6 7 8   Year: 7 8 9 0 1 2 3 4 5 6

ISBN 0-02-328690-3

Preface

PURPOSE   This engineering graphics text-workbook has been developed to meet the needs of the introductory engineering graphics program. The text is comprehensive in scope, as all topics normally studied by freshman engineering students are presented. It has been the authors' goal to discuss concisely the fundamental concepts involved–without going into lengthy, verbose discussions–in order to save both the student and instructor valuable classroom time. We have found that many complex concepts can be reduced, clarified, and stated in common, straightforward language that the engineering student can easily understand and assimilate.

CONTENTS   The text and problems cover the subjects of engineering drawing, descriptive geometry, and graphical mathematics as one integrated course. Emphasis has been placed on the fundamentals of each area of instruction, with the anticipation that each topic will be further amplified and illustrated by the instructor. The authors believe that this approach will be useful and appealing to the beginning student.

NEW TO THIS EDITION   In response to suggestions from the many users of this text, a section on computer graphics has been added to Chapter 2, stressing the use of the computer as a drawing tool. The number of problems has been reduced by eliminating those problems that duplicated basic principles. The practical application of the text material has been retained by revising more than half of the remaining problems. Problem sheets are arranged in sequence with the text material, as in previous editions. These changes have reduced the size and cost of the text, making it more practical for adoption in the shorter graphics courses of contemporary engineering programs. A complete set of solutions is available from the publisher.

ACKNOWLEDGMENTS   The authors express their appreciation to Clarice Williams, Cindy Koons, and Debbie Atkins for their assistance in the preparation of this text.

J.B.D., W.G.D., F.F.M., and H.F.T
Division of Engineering Fundamentals
Virginia Polytechnic Institute and State University

Contents

5   INTERSECTION AND DEVELOPMENTS   69

6   TECHNICAL PRACTICES   85

7 ENGINEERING DRAWINGS   115

8 VECTORS   123

9 CHARTS AND GRAPHS   129

Lettering / Sketching / Orthographic and Pictorial Views

Engineering graphics is a prime means of communication and a medium for the development of design ideas. The ability to letter and sketch is the hallmark of the competent engineer. This first chapter is devoted to a description of techniques necessary to develop the ability to work freehand with pencil and paper; to letter; and to illustrate by using orthographic and pictorial projection methods.

1.1 LETTERING

Illegible or poorly executed notes and dimensions tend to defeat the purpose of a drawing. The ability to letter neatly and rapidly can be acquired by anyone with the will to practice, but practice in making incorrectly shaped letters is of no value. The student should learn the shapes and proportions of the single stroke commercial Gothic alphabet and numerals as illustrated for the vertical style in Figure 1-1.

1.2 LETTERING TECHNIQUE

The experienced engineer will modify the letter shapes and develop a personal style, but the beginner should strive for the correct shape and proportion of each letter.

Figure 1-1 Vertical Gothic Letters and Numerals

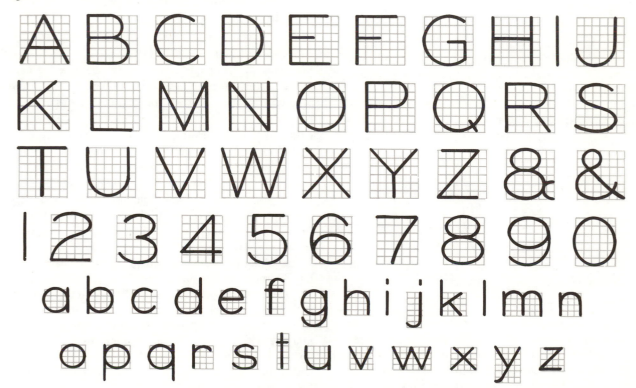

In lettering the vertical "caps" it is important to remember that the middle horizontal stroke of the letters E , F , and H is placed a little above center and that the letters B , E , K , S , X , and Z are not quite as wide on top as on the bottom; otherwise these letters would appear top-heavy. The letters I , J , and M are often formed incorrectly by the beginner.

Bold black letters are achieved by using the F grade of lead. The point should be conical in shape and at least 6 mm in length. The single strokes forming the letters are made with a finger and wrist motion. The sequence of strokes, in general, should be from top to bottom and from left to right. The pencil must be kept sharp and should be rotated frequently in the fingers to keep the width of lines uniform. The forearm should always rest on the drawing surface. Letters larger than 6 mm in height may first be sketched with light overlapping strokes, cleaned up with an eraser, and then darkened with firm single strokes.

In combining letters to form words, the letters are _not_ spaced equal distances apart—the spaces are proportioned to give equal clear areas between letters. Words should be spaced the width of the letter "O" apart.

Lettering may also be done mechanically by using special tools such as the Leroy or Wrico lettering sets, alphabet templates, lettering typewriters, or transfer letters.

1.3 GUIDELINES

Guidelines should be used for lettered notes and titles on drawings. Special devices are useful for drawing guidelines. For example, the disc-type instrument illustrated in Figure 1-2 is set for drawing horizontal guidelines for 6-mm letters.

Figure 1-2   Lettering Guide

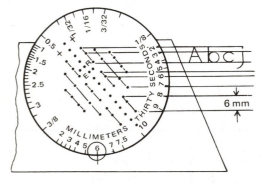

Different heights (vertical distances) between the holes are obtained by turning the disc and aligning the desired millimeter line with the index mark on the base. Guidelines are drawn by placing the sharp point of a hard lead pencil in the holes and moving the device back and forth

along the upper edge of the T-square blade. The lower hole of each group of three locates the base line, the middle hole locates the waist line for lowercase letters, and the top hole locates the cap line. The beginner should also draw random spaced vertical guidelines to avoid sloping the letters to the right or left. Instruments of this type have several parallel sets of spaced holes for guidelines for capital and lowercase letters, and for drawing evenly spaced parallel lines for section lines, etc.

1.4 TITLES

Title blocks are designed and drawn or printed in block form on the drawing sheet with blank spaces for detailed information. Map titles are designed for symmetrical layout about a center line with letter heights of the different lines dependent upon the relative importance of each line. See Figures 7-1 and 7-12.

1.5 SKETCHING

In engineering design, the first drawings are in the form of sketches. Many sketches are drawn and redrawn before a new or improved design reaches the drawing board. Sketches may be made entirely freehand or instrument-aided by templates and straightedges.

This unit on sketching is confined to freehand work where the only materials used are pencils, a sharpener or lead pointer, an eraser, and paper. HB- or F-grade lead is used, and the point is sharpened to give the width of line desired.

1.6 SKETCHING TECHNIQUE

Sketching on grid paper is not difficult because the horizontal and vertical directions are fixed and the grid squares may be counted for correct dimensions and proportion. Horizontal lines are sketched from left to right and vertical lines from top to bottom. The forearm should rest on the tabletop and the lines are drawn in lengths to correspond to the natural swing of the fingers, hand, and wrist. Very short gaps may be left between strokes, or the strokes may join for a continuous line. The paper should be turned to the most convenient angle for ease in sketching, even 90° if necessary for long vertical lines.

Sketching on plain paper requires special care in obtaining true directions and proportional size relationships. To sketch a straight line, mark the beginning and end points and aim for the end point with the eye. First, sketch with light overlapping strokes as though brushing in a line with a watercolor brush, then correct and clean up the line with the eraser, and finally go over the line with firm, black strokes. Distances are usually estimated for correct proportions.

Figure 1–3 Sketching Circles and Ellipses

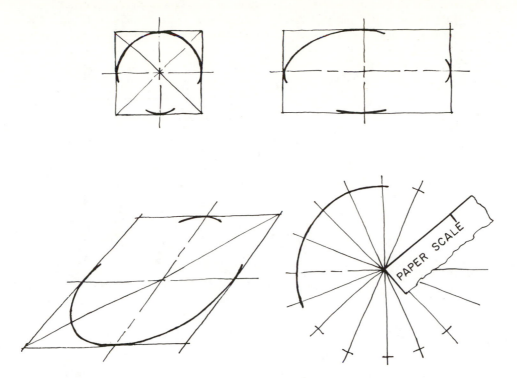

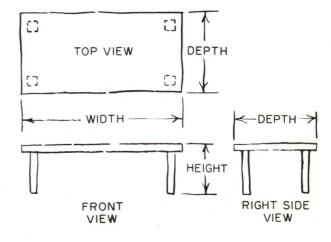

Circles and ellipses are sketched by drawing center lines and "box" lines which establish the lines of symmetry and overall dimensions of the required figure. See Figure 1–3. Four short arcs are drawn tangent to the sides of the box at the midpoint and then the missing segments are filled in. The sheet should be turned for ease in sketching each of the four arc segments.

Large circles are sketched through points set off on diameter lines drawn through the center points.

1.7 ORTHOGRAPHIC VIEWS

Three-dimensional objects may be represented on a sheet of paper by sketching one or more orthographic views. Orthographic views are obtained by looking squarely at one or more faces, as required, to describe the shape and proportions of the object. The true shape of a gasket or template can be shown by one view with a note to give the thickness of the material. A minimum of two views is normally required to describe the shape and show the width, height, and depth dimensions of an object. Theoretically, each view is projected onto a plane of projection by projecting each point of the object to the plane with projecting lines or projectors constructed perpendicular to the plane.

Since the views are projections of the same fixed object on mutually perpendicular planes, the front, top, and right-side views share exact relationships in shape, size, and position. This relationship must be retained when the views are sketched, and it is of the *utmost* importance that the top view be placed directly above the front view and the right-side view placed to the right and aligned with the front view. See Figure 1–4.

Figure 1–4 Three Orthographic Views of a Table

In order to learn to "read" engineering drawings, the student must develop the ability to visualize three-dimensional spatial relationships. Therefore, it is very important always to think of the related views and dimensions as seen from different positions in space and not as independent drawings on a single plane.

3

1.8 SKETCHING VIEWS

The correct procedure for sketching the three principal views—front, top, and right side—is to lay out the overall width, height, and depth dimensions of the object to form three rectangles to enclose the three views as shown in Figure 1-5. The front view will show width and height, the top view width and depth, and the side view height and depth.

After "blocking in" the views, the details of the geometric shapes are sketched in light lines. Details are projected back and forth from view to view. Thus all views are completed together, not one at a time.

Finally, after checking for errors and omissions, the lines are made sharp and black. All views should show all features complete with correct line symbols. See Figure 1-6.

Figure 1-7 illustrates four basic geometric shapes pictorially and by orthographic views with the correct line symbols for visible, hidden, and center lines. A sphere appears as a circle in all views. All designs include composites of the basic geometric shapes.

Figure 1-5 Blocking in Views

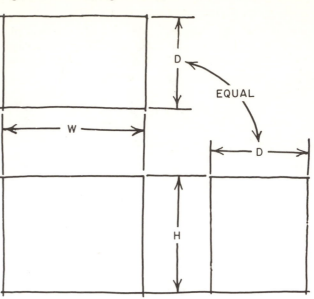

Figure 1-6 Line Symbols

Figure 1-7 Four Basic Geometric Shapes

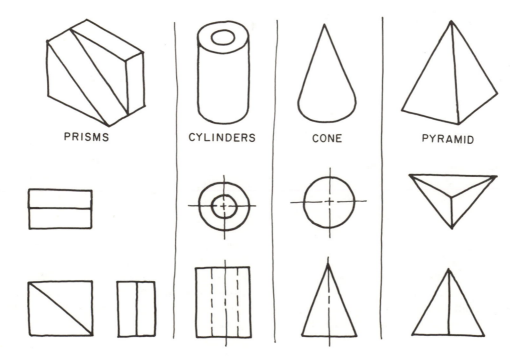

4

1.9 ISOMETRIC PROJECTION

An orthographic view will be a three-dimensional pictorial if the principal faces of the object are inclined to the plane of projection. Single-view pictorials within the orthographic system are classified as isometric, dimetric, and trimetric under the general heading of axonometric projection.

In *dimetric projection*, only two of the three principal axes (width, depth, and height) make equal angles with the plane of projection.

In *trimetric projection*, the three principal axes make different angles with the plane of projection.

Three different scales are used for a trimetric drawing, two for dimetric, but only one scale for isometric drawing.

Isometric is the easiest of the three to draw and is therefore most often used. A comparison of a normal front view of a cube with an isometric is shown pictorially in Figure 1-8.

Figure 1-8 Orthographic Projection of a Cube

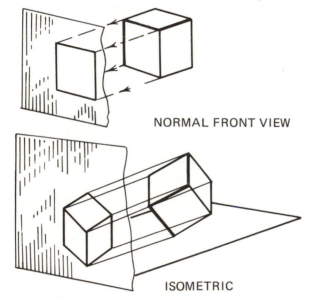

NORMAL FRONT VIEW

ISOMETRIC

In *isometric projection* the three principal axes of the object make equal angles (35°16′) with the plane of projection. A cube in this position would have a rear corner directly behind the front corner and the body diagonal from front to rear would be parallel to the line of sight and therefore perpendicular to the projection plane. The projections of the axes or edges of the cube will form angles of 120° with each other and the image of the cube on the plane will be smaller than the actual cube in the ratio of 0.816 to 1. To avoid using a special isometric scale, the reduction is usually ignored, and the drawing is made in a ratio of 1:1. This is called an *isometric drawing*, as distinguished from the true isometric projection of an object. See Figure 1-9.

Figure 1-9 Isometric Projection and Isometric Drawing of a 1 unit Cube

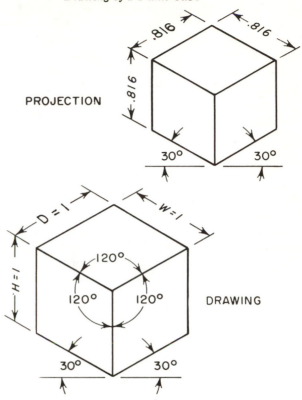

PROJECTION

DRAWING

The projections of the principal axes on the drawing are called the *isometric axes*, and the projections of all lines parallel to the principal axes are called *isometric lines*.

1.10 ISOMETRIC SKETCHING

Isometric sketches are made by sketching the isometric axes at angles of 120° with each other. The height axis is usually vertical and the other two up to the right and left at 30° to the horizontal as shown in Figure 1-9. Overall width, depth, and height distances are measured on the isometric axes, and the "box" that will enclose the view is constructed by drawing parallel edge lines. The view is completed by drawing all visible lines. Hidden lines are not shown unless essential for clarity.

Measurements can only be made on isometric lines.

1.11 COORDINATE CONSTRUCTION

Nonisometric lines are drawn by locating points on the line by coordinates. The 30° and 45° angle lines in Figure 1-10 are located by transferring the coordinate distances *a, b,* and *c* from the axes on the true-size view to the isometric axes. Irregularly shaped objects are drawn by offset or coordinate construction. Points are located as illustrated in Figure 1-11 for point 1 on the circular arc and point 1′ on the elliptical curve in the nonisometric plane.

Figure 1-10 Angles by Coordinates

Figure 1-11 Coordinate Construction

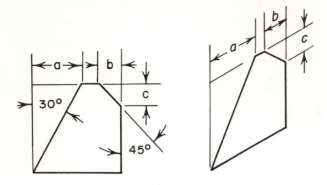

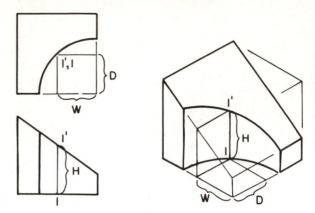

1.12 CIRCLES IN ISOMETRIC

A circle will project as an ellipse when the plane of the circle is inclined to the plane of projection. An ellipse in an isometric plane is sketched by first sketching the rhombus, which is the projection of the square that circumscribes the circle. See Figures 1-3 and 1-12. The construction for drawing an approximate ellipse with the compass is illustrated in Figure 2-13.

A cylinder is sketched by constructing the two elliptical bases and then drawing object lines parallel to the cylinder axis and tangent to the base curves at the points of intersection with the long rhombus diagonals. The long diagonals will always be perpendicular to the axis of the cylinder.

Figure 1-12 Sketching Ellipses in Isometric

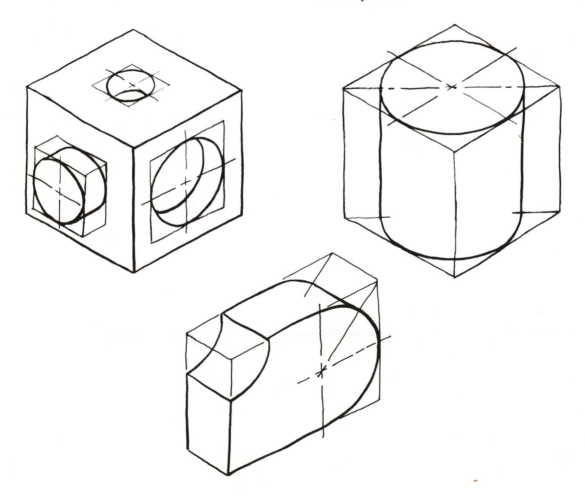

Figure 1-13   Orthographic and Oblique Projection

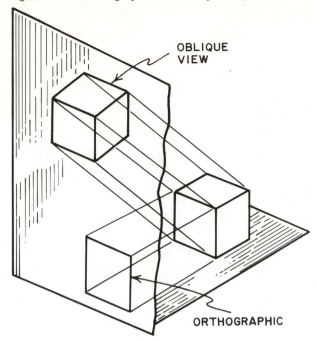

1.13   OBLIQUE PROJECTION

Cavalier, cabinet, and *general* oblique drawings are single-view pictorial projections in which the projecting lines are parallel to each other and oblique (not perpendicular) to the plane of projection. See Figure 1-13.

The front face and all elements parallel to the frontal plane will project <u>true shape and size</u>. Therefore, the width and height axes are sketched in their normal position.

The depth axis, called the *receding axis*, is drawn at any desired angle, θ, with the horizontal and is measured in any ratio in comparison with the height and width axes.

If the oblique *projecting lines* are at an angle of 45° with the plane of projection, the ratio will be 1 to 1 and the view is a *cavalier* projection. If the corresponding ratio is 1 to ½, the view is a *cabinet* drawing. All other

ratios are classified as *general oblique*. In Figure 1-14, θ may be any desired angle.

1.14   OBLIQUE SKETCHING

In sketching an oblique view, the angle, direction, and scale of the receding axis are selected to give maximum definition and least distortion. The axis is usually drawn at either 30° or 45° upward to the right or left. A cabinet or general oblique will reduce distortion by decreasing the depth dimension.

In general, the long dimension of the object should be selected as width, but not if this position places the most irregular features in a receding plane.

<u>*Measurements in the receding planes can only be made parallel to the axes.*</u> Particular care must be exercised when two scales are used, as in cabinet and general oblique drawings.

Angular cuts and curves in inclined planes are located by offset measurements or coordinates the same as in isometric drawings.

Circles will project as ellipses in the receding planes and are sketched as illustrated in Figure 1-15.

Figure 1-15   Sketching Circles and Ellipses in Oblique

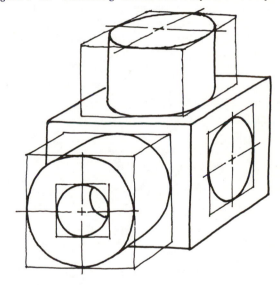

Figure 1-14   Types of Oblique Projection

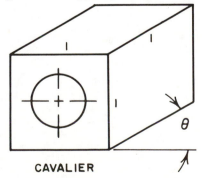

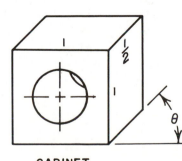

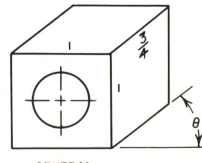

CAVALIER   CABINET   GENERAL

Figure 1-16 The Perspective System

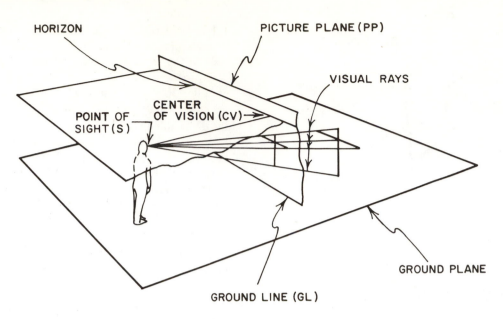

1.15 PERSPECTIVE

Perspective is that form of projection in which the projecting lines or visual rays radiate from a point of sight located a finite distance from the object. It produces the most realistic image of all forms of projection. The nomenclature and space relations of the elements of a perspective system are illustrated in Figure 1-16.

Skill in making freehand perspectives requires study and practice but pseudoperspectives are easily made and may be useful to the engineer whenever an isometric or oblique sketch appears too distorted. The view is sketched in a manner similar to isometric or oblique with one major change. The receding parallel horizontal line systems are drawn to converge at vanishing points on a horizon line. The vanishing points are selected to give the desired view and the receding planes are foreshortened visually for optimum effect. See Figure 1-17.

Figure 1-17 Perspective Sketching

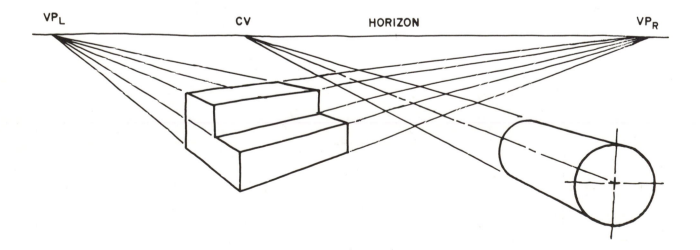

Figure 1-18 Top-View Layout for Angular Perspective of a Prism

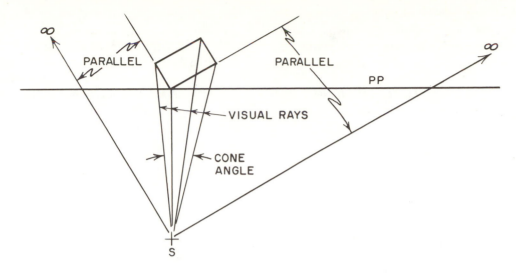

1.16 ANGULAR PERSPECTIVE

Mechanical perspectives are made by drawing ortho-graphic views of the system. The *angular* or *two-point perspective* of a rectangular prism is constructed as shown in Figures 1-18 and 1-19. The object is normally placed with a front corner in the picture plane and with the front face at 30° with the plane. The point of sight is selected to give the desired view with the least distortion. The cone of vision angle should not exceed 30°.

The top view or plan (Figure 1-18) is drawn first to show the relative positions of the object, the picture plane (*PP*), the point of sight (*S*), the visual rays, and the construction lines through *S* to infinity (∞).

A front view of the picture plane will show the per-spective of the prism as determined by the points in which the visual rays pierce the plane. See Figure 1-19.

The horizon line is drawn at eye height above the ground line at any convenient distance from the top view. A front-elevation view of the prism is drawn for convenience in projecting height dimensions.

The perspective of the front corner *AB* will show true height because it lies in the picture plane. Points *A* and *B* are projected from the top view and the height from the side view. The vertical edge *CD* will not show true height. *Height can only be measured in the picture plane.* Therefore, points *C* and *D* are projected into the picture plane to points *A* and *B* and/or to points *C'* and *D'*. The horizontal edge lines of the prism are then drawn to the correct vanishing points VP_L and VP_R on the horizon, and the vertical edges projected from the top view to complete the perspective.

Figure 1-19 Angular Perspective of a Prism

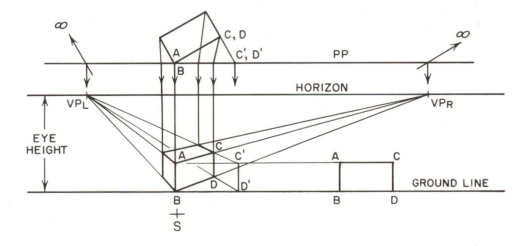

1.17 PARALLEL PERSPECTIVE

In *parallel* or *one-point perspective* the object is placed the same as in oblique projection, with the principal contour surfaces parallel to the picture plane. In this position the height and width axes will project in true position and the depth axis will vanish at the center of vision (*CV*).

In Figure 1-20, the front face of the cylinder, which is behind the picture plane, has been moved forward into the plane for measurement. The front view is then drawn for this position and the cone formed by the contour lines drawn to the center of vision will be the perspective of the cylinder infinite in length. The front and back surfaces of the object are located by visual rays from the point of sight. The circles will project as true circles with decreasing radii.

Figure 1-20 Parallel Perspective

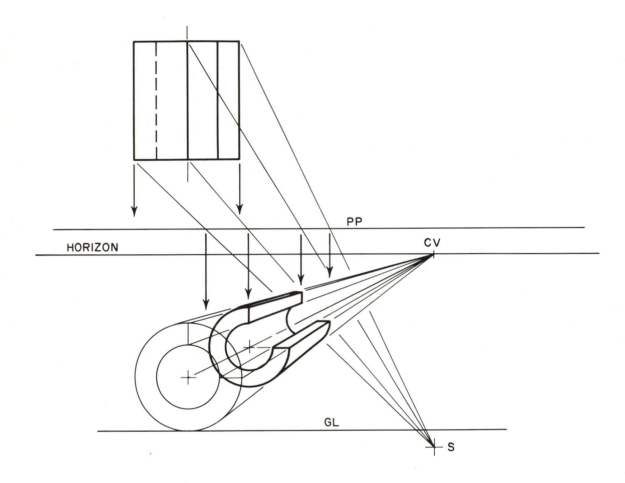

Drawing Equipment / Geometric Construction

2

The basic drawing equipment for an engineer consists of paper, drafting tape, pencils, sharpener, erasers, erasing shield, drawing board, T-square, triangles, scales, irregular curves, compass, dividers, and protractor.

Special equipment includes drafting machines, templates, adjustable curves, ruling pens, proportional dividers, beam compass, electric eraser, railroad and ship curves, planimeter and other items illustrated in manufacturers' catalogues.

2.1 PAPER AND PENCILS

Drawing paper is available in numerous grades, colors, and standard sizes to satisfy industrial requirements. Final working drawings are usually made in either pencil or ink on transparent paper (vellum), tracing cloth, or polyester film (Mylar).

The correct grade of lead to use depends upon the paper and the type of line desired. Lead holders are made for standard-size leads and mechanical pencils for 0.7 mm, 0.5 mm, and 0.3 mm diameter leads. The standard leads range in hardness from a very hard 9H to medium H, F, HB, and B, to very soft 6B. Standard leads are sharpened with a lead pointer to obtain the correct line width from a wide object line to a very light construction line. The thin leads are designed for a uniform line width and are not sharpened.

Erasing is done with a soft nongrit eraser that must be kept clean and dry. An erasing shield is used to protect adjacent lines. General cleaning is done with a soft Artgum eraser. To avoid dirty, smudged drawings, the equipment, particularly the triangles and T-square, must be kept clean.

2.2 T-SQUARE AND TRIANGLES

The T-square and the 45° and 30°-60° triangles are used to draw straight lines. The drawing sheet is fastened in position with drafting tape with the left edge placed close to the left edge of the board and with the horizontal border line aligned with the T-square blade.

Horizontal lines are drawn from left to right along the upper edge of the T-square blade, and vertical lines are drawn from the bottom toward the top of the sheet along the left vertical edge of a triangle. See Figure 2-1.

Figure 2-1 Horizontal and Vertical Lines

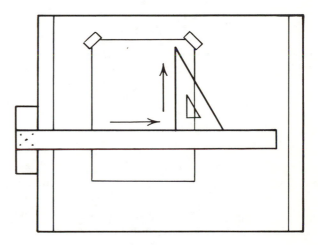

The 45° and 30°-60° triangles may be used in various combinations with the T-square to draw lines at 15°, 30°, 45°, 60°, and 75° with a given line. See Figure 2-2.

The protractor is used to locate lines at angles that are not multiples of 15°. For accurate work both the 0° base line and the 90° line should be placed in alignment with two perpendicular lines constructed through the point of origin.

The triangles are also used to draw lines parallel or perpendicular to a given line. In Figure 2-3 one leg of the triangle is aligned with the given line and the triangle moved to the dashed position by sliding it along the supporting edge of the other triangle or T-square blade.

Figure 2-2 15° and 75° Lines

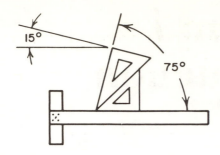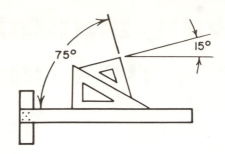

Figure 2-3 Parallel and Perpendicular Lines

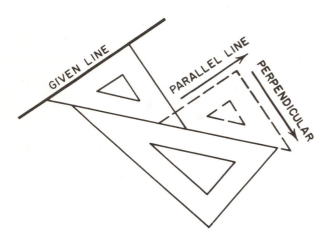

2.3 SCALES

A scale is a device used to make measurements on engineering drawings. The metric and engineer's scales are divided in decimal divisions. The architect's scale is divided in common fraction divisions of the inch.

The commonly used triangular metric scale is 30 cm long. Each side has two divided scales, one reading from the right and one from the left. Since the meter is the basic unit of linear measurement in the metric system,

there is no reference to English units on the metric scale. All numbered divisions on the scale correspond to meters at the given *scale ratios*. The most common scale ratios used are 1:100, 1:80, 1:50, 1:40, 1:33⅓, and 1:20. Corresponding *scale factors* are shown with each scale ratio, to indicate what decimal part of a meter each numbered division represents.

Thus, on the 1:40 scale shown in Figure 2-4, the 1:40 ratio indicates that one unit of length on the scale or drawing represents 40 such units on the object. The 0.025 scale factor indicates that each *numbered* division on the scale is actually 0.025 m long.

Using the 1:40 metric scale and a ratio of 1:40, the line in Figure 2-4 is measured as 5.54 meters (m) or 554 centimeters (cm). If the scale of the drawing were 1:400, the same line would be measured as 55.4 m, since the drawing scale is 10 times the scale ratio.

The 1:100 metric scale may be used for full-size measurements in centimeters, since the scale factor indicates that each numbered unit length is .01 of a meter, or 1 centimeter. Thus the line in Figure 2-5 is measured as 9.7 cm or 97 millimeters (mm) at a drawing scale of 1:1 (full size). If the drawing scale were 1:10 000, the line would represent 970 meters, since the drawing scale is 100 times the 1:100 scale ratio.

Figure 2-4 Metric Scale

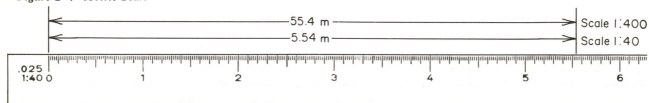

Figure 2-5 Metric Scale

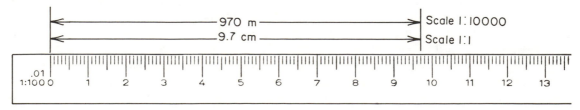

Figure 2-6 Engineer's Scale

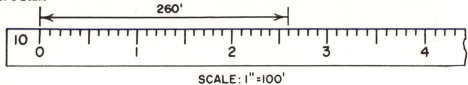

SCALE: 1"=100'

Figure 2-7 Architect's Scale

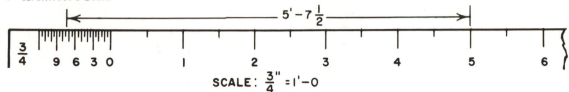

SCALE: $\frac{3}{4}$" = 1'-0

The triangular engineer's scale has fully divided inches on each face, with the identifying numbers 10, 20, 30, etc., indicating the number of divisions per inch. This scale may be used for measuring any quantities in decimal divisions based on the English foot or inch. Figure 2-6 illustrates the 10 scale, which is used for scales such as 1" = 10', 1" = 100 yds., and 1" = 1000 meters. The actual quantity measure represented depends upon the selected scale of the drawing.

The 12" triangular architect's scale has one face marked 16. This standard foot distance has each inch subdivided into sixteenths. It is used to make full-size, half-size, and enlarged drawings when the English foot is the basic unit. The other faces are designed for making reduced-size drawings from 1/4 size (3" = 1'-0) to 1/128 size (3/32" = 1'-0). Each face has two open divided scales, one reading from the right and one from the left. The scale marked 3/4 (3/4" = 1'-0) has 14 numbered 3/4" divisions to the right of the 0 mark, each represent-ing 1 foot, and one 3/4" division to the left of the 0 mark, subdivided into inches and half-inches. Foot distances are measured to the right and inches and half-inches to the left of the 0 mark, with one position of the scale. See Figure 2-7. The other scale on this face is 1/32 size (3/8" = 1'-0), with 3/8" (foot) divisions reading from the right end of the scale. The other faces also have scales designed with the foot represented by some fractional value of a standard foot. In each case, the standard foot distance has simply been compressed into a shorter distance, with proportionally shorter inches and fractions of inches.

Scaled distances are compared in Table 2-1, page 14.

2.4 COMPASS

The compass is used to draw circles and arcs and may also be used to transfer distances. Accurate compass work requires correct adjustment of the legs. The shoulder end of the needle point should be set slightly longer than the sharpened lead point. The lead should project 6 mm and be sharpened with a file or sandpaper to a wedge shape, as illustrated in Figure 2-8.

Circle templates are often used to avoid tedious compass work, particularly in drawing rounded corners on machine drawings and rivet symbols in structural drawings.

Figure 2-8 Compass Points

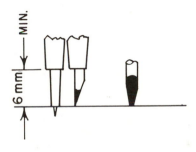

2.5 DIVIDERS

The dividers are used to set off or transfer distances and to divide a line into equal parts. Distances should first be measured on a line with the scale and then transferred with the dividers.

To divide a line *AB* into three equal divisions (Figure 2-9), a trial division is made with the dividers set to an estimated one third of the length. The setting is then corrected by one third of the error *E* and another trial division stepped off. The second or third trial will normally result in an accurate division of the line.

Figure 2-9 Line Division with Dividers

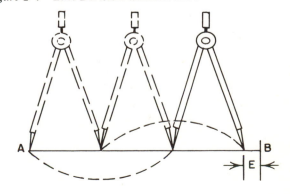

Table 2-1 Use of Scales

The actual length of this line is 0.203 m or 8 in.

	Scale Factor	Drawing Scale	Scale Ratio	Distance Represented by Line Above
METRIC SCALE (30 cm long for convenience)	.01	1:100 1:1	1:100	20.3 m 0.203 m or 20.3 cm
	.05	1:20 1:2	1:20	4.06 m 0.406 m or 40.6 cm
	.0125	1:80 1:8 1:800	1:80	16.24 m 1.624 m or 162.4 cm 162.4 m
ENGINEER'S SCALE (12" long for convenience) 1 inch = "X" units === (key)	**Scale** 10	1" = 1' 1" = 10'	1:12 1:120	8' 80'
	30	1" = 30" 1" = 3' 1" = 30'	1:30 1:36 1:360	240" = 20' 24' 240'
	50	1" = 50' 1" = 5'	1:600 1:60	400' 40'
ARCHITECT'S SCALE (12" long for convenience) "X" inches = 1 foot === (key)	1	1" = 1'-0	1:12	8'-0
	$1\frac{1}{2}$	$1\frac{1}{2}$" = 1'-0	1:8	5'-4
	3/8	3/8" = 1'-0	1:32	21'-4

Figure 2-10   French Curve

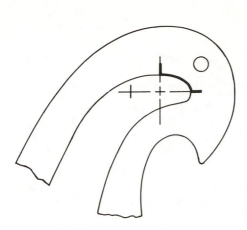 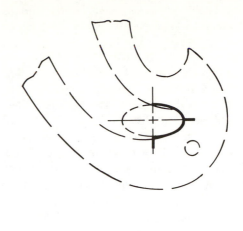

2.6   FRENCH CURVES

French curves, also known as *irregular curves,* are made in a large variety of shapes and sizes. They are used in drawing noncircular curves, usually through a series of plotted points. To draw a curve, first sketch a light free-hand line through the points. Then draw the curve by fitting selected portions of the French curve to the free-hand line. Successive positions of the curve must overlap and be tangent to each other to assure a smooth, faired line—a curve without irregularities or breaks.

Small ellipses can be drawn with four positions of the correct curve segment. The quarter sections are made identical by marking the edge of the curve with a soft pencil. In Figure 2-10, the solid-line French curve is marked in position to draw the upper-right quarter of an ellipse. The same part of the curve is then used to draw the lower right quarter by turning the curve over and fitting the marks to the axis lines.

2.7   ENGINEERING GRAPHICS AND THE DIGITAL COMPUTER

The digital computer is a powerful and versatile instrument to aid in the preparation of engineering drawings. It is, in effect, a modern drawing instrument. As the cost of hardware and software has steadily declined, this new tool has experienced increasing acceptance and use in the graphics field.

The term *computer graphics* encompasses all types of computer drawing from simple line, bar, and pie charts to complete engineering drawings as well as the fantastic animated art now commonly viewed on television screens. In the engineering profession, acronyms commonly used are CADD (computer-aided design and drafting) and for the follow-on process, CAM (computer-aided manufacturing).

It is most important to remember, however, that the digital computer and its peripheral devices are machines. The computer cannot create; it can only do what it is instructed to do. It is therefore imperative that the designer or drafter have a complete understanding of the principles and techniques of engineering graphics in order to use the tool effectively. This is why "aided" is attached to "computer" in CADD or CAM.

Competition in the marketplace has driven down the cost of both computer hardware and the software needed to make the hardware perform. Nevertheless, computer graphics systems remain relatively expensive. The development of sophisticated computer graphics software entails thousands of man-hours of skilled programming effort. The prospective purchaser/user must carefully weigh the benefits to be derived from such systems against the initial and maintenance costs involved.

2.8   GRAPHICS WITH THE COMPUTER

The graphics output of a digital computer is generally a display on the face of a cathode ray tube (monitor) and/or a drawing produced by an attached plotter or printer. In any case, the product is a two-dimensional display of a series of straight lines. The computer can draw a straight line from one point to another. Arcs are made up of small straight line segments; the shorter the segments, the more closely the displayed arc approaches the true arc. Pictorial (3D) displays call for an additional depth axis and, in most cases, mathematical computation to locate properly all points of an object on the display. The mathematics generally involves substantial vector and matrix manipulations that require extensive computer memory capacity.

Computer monitor *resolution* is a factor of the number of addressable points or screen *pixels*. Turning a pixel "on" creates a visible point. Lines are made up of contiguous "on" pixels. The higher the resolution, the "finer" the display. The IBM Color Monitor has both a medium and a high resolution capability. The medium resolution screen is 320 pixels wide by 200 pixels high. High resolution is 640 × 200. Since there are more pixels per unit distance in the X (width) dimension than in the Y (height) dimension, a scaling problem is inherent in either mode and must be considered in developing the display. Some modern *work stations* devoted to computer graphics and design utilize a very high resolution 1024 × 1024 pixel display, which eliminates this scaling problem. A complete work-station might include one or more input devices (keyboard, digitizer pad, mouse, light pen), its own dedicated minicomputer (or expanded memory microcomputer), and output devices (display monitor, plotter, printer).

Graphics software is readily available from many commercial sources. Software packages vary considerably in quality, capability, machine dependence, and "user friendliness." Users should certainly make sure the software meets their drawing/design criteria before investing. Users may also wish to write their own graphics programs. BASICA is a commonly used graphics programming language, as is FORTRAN in conjunction with PLOT 10.

The BASICA program in Figure 2-11 will produce a display of the construction to determine the true length of a line. The actual true length and the angle of the line with the horizontal plane are computed mathematically. Figure 2-12 is a printout of results from this program in medium resolution. While the graphic output is similar to a pencil and paper construction, the limitations of the output display preclude accurate measurements from the display itself. Minor changes to the same BASICA program (high resolution screen, scale factors) produced the output seen in Figure 2-13.

Figure 2-11 BASICA Program—True Length of a Line

```
10 REM          BASICA Program -- True Length of a Line
20 REM
30 CLS : KEY OFF
40 SCREEN 1,0                  ' Medium resolution screen, no color
50 D=20                 ' Arbitrary distance from H view of line to H/1 fold line
60 X0=50: Y0=108        ' Position of origin - left end of H/F fold line
70 X1=180: Y1=167       ' Bottom of F/P fold line
80 REM          Stmts. 100 through 150 establish plot area and annotations
90 REM
100 LINE (X0-20,Y0)-(X1,Y0),,,&HFFCC           ' H/F fold line
110 LINE(X1,Y0)-(X1,Y1),,,&HFFCC           ' F/P fold line
120 LOCATE 12,40: PRINT "X": LOCATE 14,36: PRINT "Y":LOCATE 10,38: PRINT "Z"
130 LINE (284,104)-(284,92): LINE-(298,80): LINE (284,92)-(300,92)
140 LOCATE 13,22: PRINT "H" : LOCATE 15,22 : PRINT "F"
150 LOCATE 15,1: PRINT "(0,0,0)" : LOCATE 15,24 : PRINT "P"
160 REM
170 REM     Input is position of points A and B in the F,P, and H views. In
180 REM     other words, the X, Y, and Z (WIDTH, HEIGHT, DEPTH) coordinates.
190 REM     >>>> X range is 30 to 180, Y is 0 to 40, Z is 0 to 40 <<<<
200 LOCATE 23,1 : INPUT"XA,YA,ZA,XB,YB,ZB";XA,YA,ZA,XB,YB,ZB
210 REM
220 REM     Stmts. 230 through 250 establish location of A and B in F,P, and H
230 AXF=X0 + XA : AYF=Y0 + YA : BXF=X0 + XB : BYF=Y0 + YB
240 AXP=X1 + ZA * 1.2 : AYP=AYF : BXP=X1 + ZB * 1.2 : BYP=BYF
250 AXH=AXF : AYH=Y0 - ZA : BXH=BXF : BYH=Y0 -ZB
260 REM
270 REM          'Screen Scale Factor, horizontal / vertical, is 1.2
280 REM
290 ALPHA=ATN((BYH-AYH)/(BXH-AXH)) ' Angle of line in H with H/F fold line
300 REM
310 REM     C and S are legs of hypotenuse distance D to plot H/1 fold line
320 C= D * COS(ALPHA): S= D * SIN(ALPHA) * 1.2
330 REM
340 FL1X=AXH+S: FL1Y=AYH-C          'FL1,2,3,4 are points on H/1 fold line
350 FL2X=FL1X-C: FL2Y=FL1Y -S *5/6
360 FL3X= BXH+S: FL3Y= BYH - C
370 FL4X= FL3X+C: FL4Y = FL3Y + S *5/6
380 REM          Stmts. 390-400 compute A and B in Auxiliary View 1
390 AXA= AXH +((D +YA) * SIN(ALPHA)) * 1.2 : AYA= AYH -(D + YA) * COS(ALPHA)
400 BXA= BXH +((D +YB) * SIN(ALPHA)) * 1.2 : BYA= BYH -(D + YB) * COS(ALPHA)
410 REM
420 REM     We may now draw the H/1 fold line and line AB in all views
430 REM
440 LINE(FL2X,FL2Y)-(FL4X,FL4Y),,,&HFFCC           ' H/1 fold line
450 LINE(AXF,AYF)-(BXF,BYF)                ' front
460 LINE(AXP,AYP)-(BXP,BYP)                ' profile
470 LINE(AXH,AYH)-(BXH,BYH)                ' horizontal
480 LINE(AXA,AYA)-(BXA,BYA)                ' line AB in auxiliary view
490 AROW=AYF/8: ACOL=AXF/8-1: BROW=BYF/8+1: BCOL=BXF/8+2
500 LOCATE AROW,ACOL: PRINT "A": LOCATE BROW,BCOL: PRINT "B"
510 LHROW=(FL2Y/8)+2 : LHCOL=(FL2X/8)+1: LOCATE LHROW,LHCOL: PRINT "H"
520 N1ROW=(FL2Y/8)-1 : N1COL=(FL2X/8)+1 : LOCATE N1ROW,N1COL: PRINT "1"
530 TROW=(BYA+AYA)/16 - 1 : TCOL=(BXA+AXA)/16:LOCATE TROW,TCOL:PRINT "TL"
540 T1 = SQR((ZA -ZB )^2 + (XA -XB )^2)    'Math computation of True Length
550 T2 = YA -YB
560 TL = SQR(T1^2 + T2^2)
570 THETA = ATN(T2/T1)*180!/3.141593       ' Computes angle of line AB with H
580 LOCATE 24,1: PRINT USING "TL= ##.##"; TL ;: PRINT USING "   ANGLE WITH H= +#
#.## DEGREES";THETA
590 END
```

Figure 2-12 True Length of a Line—
                  Medium Resolution Screen

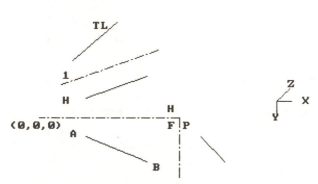

XA,YA,ZA,XB,YB,ZB? 30,18,19,95,42,40
TL= 72.40   ANGLE WITH H= -19.36 DEGREES

Figure 2-13 True Length of a Line—
                  High Resolution Screen

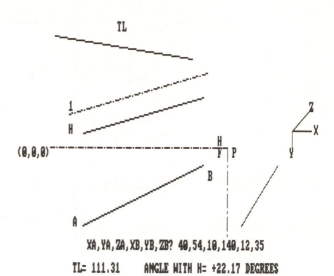

XA,YA,ZA,XB,YB,ZB? 40,54,10,140,12,35
TL= 111.31   ANGLE WITH H= +22.17 DEGREES

Examples of output from various software packages are depicted in Figures 2-14 through 2-19. All except Figure 2-19 were produced using an IBM PC (640K) and a plotter or graphics printer. A minicomputer was used for Figure 2-19. Software credits accompany each figure.

Figure 2-14 Drawings from a Printer—FIRSTDRAW 2.0

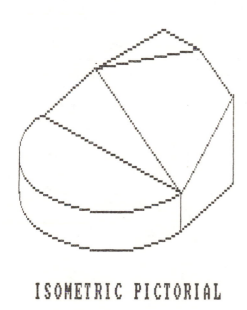

ISOMETRIC PICTORIAL

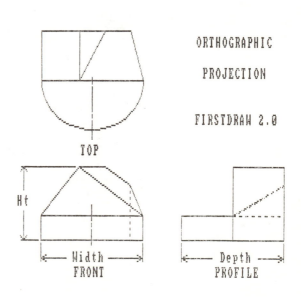

ORTHOGRAPHIC

PROJECTION

FIRSTDRAW 2.0

TOP

Ht

Width
FRONT

Depth
PROFILE

Figure 2-15 Printed Output—Interactive Microcomputer Graphics

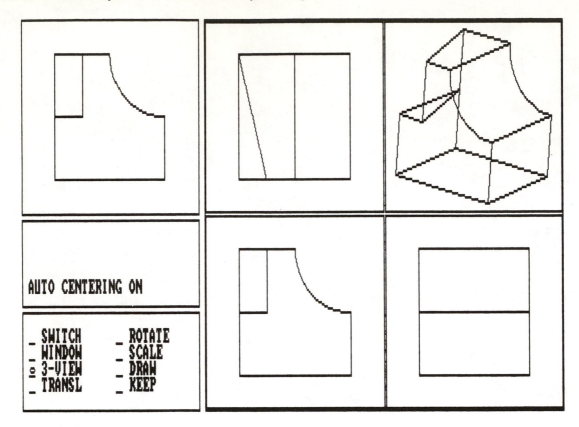

Figure 2-16 Printer Drawing—PRODESIGN II

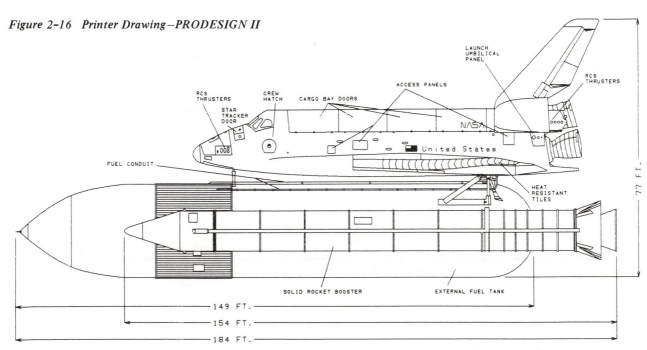

THE SPACE SHUTTLE DISCOVERY

Courtesy of American Small Business Computers

Figure 2-17   AutoCAD Plotter Drawing

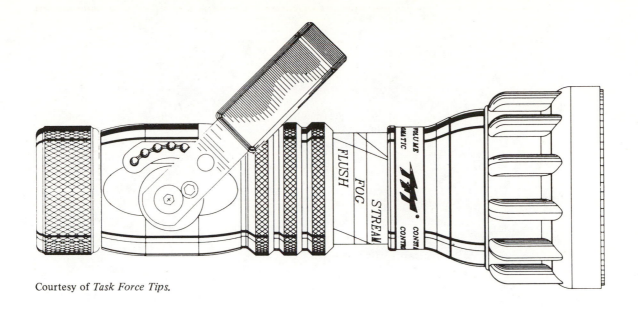

Courtesy of *Task Force Tips.*

Figure 2-18   Computer-Generated Drawings—VersaCAD, T&W Systems

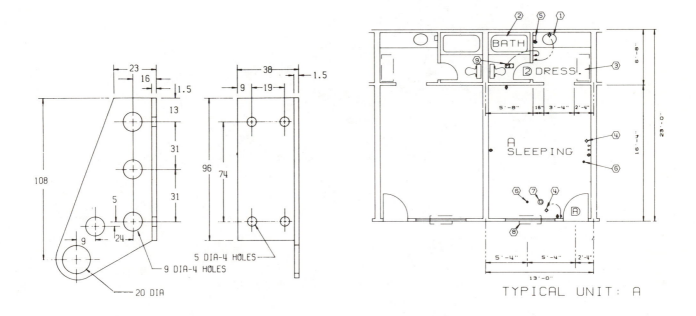

Figure 2-19   Three-Dimensional Model–INTERGRAPH

Courtesy of Intergraph Corporation

Figure 2-20   Computer Graphics Workstation–INTERGRAPH

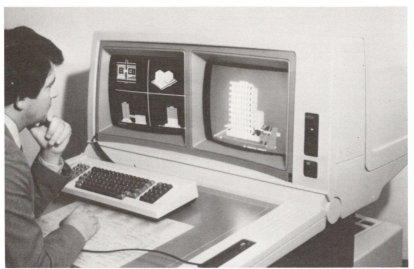

Courtesy of Intergraph Corporation

2.9 COMMENT

Modern, fast-developing computer graphics hardware and software provide the engineer with an enhanced graphics capability to aid drawing and design. The brief overview above is included in this "Drawing Equipment/Geometric Construction" chapter because the computer, in the graphics context, is a new piece of drawing equipment. Computers cannot think, visualize, or operate without instructions. More than ever before, the engineer must be able to draw upon the principles and techniques of engineering graphics for visualization of design concepts in order to efficiently use this new technology. One might add: "If you can't do it on paper, you can't do it on the computer."

References

1. "Computerized Descriptive Geometry," Dao-ning Ying, *Engineering Design Graphics Journal*, Autumn 1982, Volume 46, Number 3.
2. *Introduction to Computer Graphics*, Demel and Miller, Brooks/Cole Engineering Division, Wadsworth Inc., 1984.
3. Interactive Microcomputer Graphics, Chan S. Park, Addison-Wesley Publishing Company, 1985.

2.10 GEOMETRIC CONSTRUCTION—
CENTERS AND TANGENT POINTS

Accurate graphic constructions of applied plane geometry principles are easily made with drafting instruments. The construction lines should be drawn as light as possible.

Figure 2-21(a) illustrates the compass construction for connecting two perpendicular lines with an arc of given radius R. This construction is applicable only to perpendicular lines and locates the tangent points T before the center point C is found.

Figures 2-21(b) and (c) illustrate the construction for drawing an arc tangent to two nonperpendicular lines. The required center C is located at the intersection of two locus lines drawn parallel to and equal distance (R) from the given lines. The tangent points T are then located by constructing perpendiculars to the given lines through the center point C.

The constructions for drawing an arc of given radius R to connect a straight line and an arc or to connect two arcs are shown in Figure 2-22. The construction arc is drawn concentric with the given arc by adding or subtracting the given radius R. The tangent point of tangent arcs is located on the line connecting their respective centers.

Figure 2-21   Tangent Arcs to Straight Lines

Figure 2-22   Tangent Arcs to Curved Lines

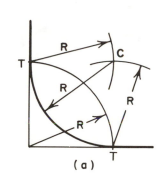

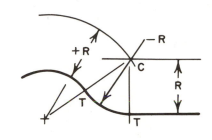

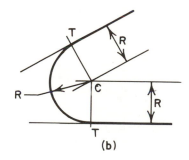

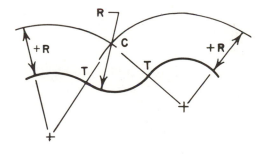

(a)

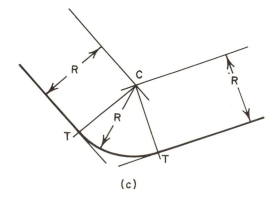

(b)

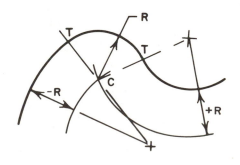

(c)

Figure 2-23 Four-Center Construction

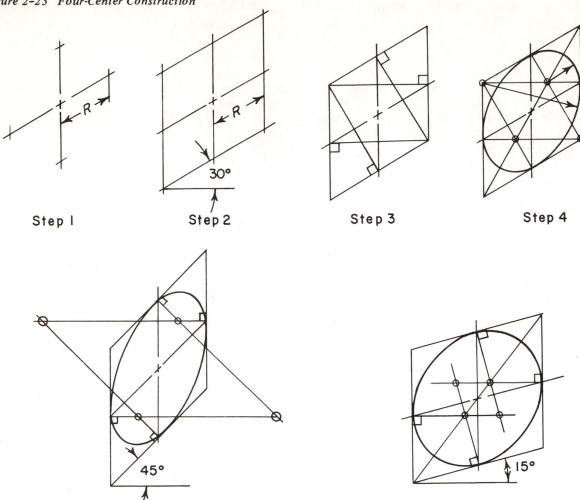

Step 1   Step 2   Step 3   Step 4

2.11 FOUR-CENTER APPROXIMATE ELLIPSE

Figure 2-23 illustrates the four-center approximate method of drawing an ellipse in a rhombus. A rhombus is a parallelogram with equal sides. This construction may be used to draw the elliptical view of a circle in isometric or cavalier oblique projection.

The centers for the four tangent arcs are located at the intersections of the *perpendicular bisectors of the sides of the rhombus.*

2.12 DIVISION OF A LINE

A line may be divided into equal or proportional parts by using the triangles to draw parallel lines (Figure 2-3). In Figure 2-24, the line *AB* is divided into five equal parts by first drawing any line through *A* and marking

off five equal distances from *A* to 5 with the dividers or scale.

Equal divisions are then located on *AB* by connecting point 5 to *B* and drawing lines *parallel* to 5–*B* through the other division points.

Figure 2-24 Line Division

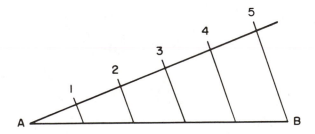

Figure 2-25 Bisectors

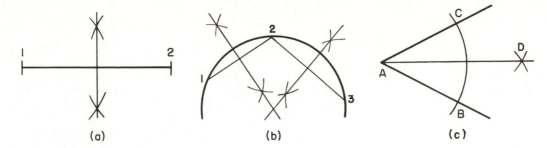

(a)   (b)   (c)

2.13 BISECTORS

The perpendicular bisector of a line is the locus of all points equidistant from the end points of the line. Therefore, intersecting arcs of equal radii, drawn with the end points as centers, will locate points on the bisector.

This construction is used to locate the center of a circle through three points by finding the point of intersection of two of the perpendicular bisectors of the lines connecting the points. See Figures 2–25(a) and (b).

The construction of an angle bisector is shown in Figure 2–25(c). Point *A* is the center of an arc of any radius drawn to locate center points *B* and *C*. Then intersecting arcs of equal radius are drawn with *B* and *C* as centers. The arcs intersect at a point *D* on the bisector.

2.14 TRANSFER OF TRIANGLES

A triangle *ABC* (Figure 2-26) may be transfered to another position by drawing one side to scale in any specified position, such as *A'B'*. Point *C* is then located by striking arcs from *A'* and *B'* with radii *AC* and *BC* respectively.

A polygon of more than three sides may be transferred by triangulation, which simply means that the polygon is divided into adjacent triangles, which are then transferred in sequence to the new position.

Figure 2-26 Transfer of a Triangle

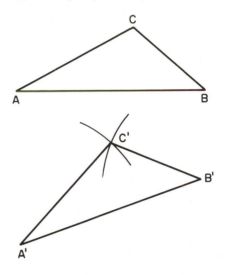

Fundamental Spatial Relationships

3

Engineering graphics is a universal language based on fundamental spatial relationships using principles of projection to establish a systematic method of representing three-dimensional objects on two dimensional displays. Several methods of projection are available to the engineer. Pictorials, including isometric, oblique, and perspective, are presented in Chapter 1. Orthographic projection is used for most engineering drawings and the graphical solution of engineering problems.

3.1 ORTHOGRAPHIC PROJECTION

Orthographic projection is the projection by parallel projectors from an object, perpendicular to a plane of projection. The most widely accepted method of describing an object is shown in Figure 3-1. An imaginary glass box with six mutually perpendicular planes is placed around the object. All points of the object are projected orthographically onto each of the six principal planes of projection. When the box is unfolded, the six *principal views* of the object will appear as shown. The relationship of views to each other is thus established. Usually the front, top, and side views are sufficient to describe the shape of an object completely.

The projection of the object onto any other plane not parallel to one of the principal planes is called an *auxiliary view* and is used to describe some feature of the object that cannot be described in a principal view.

The front plane of projection and all planes parallel to it will be called *frontal planes*. All planes parallel to the horizontal plane of projection will be called *horizontal planes,* and all planes parallel to the profile plane of projection will be called *profile planes*.

Figure 3-1 Principal Views

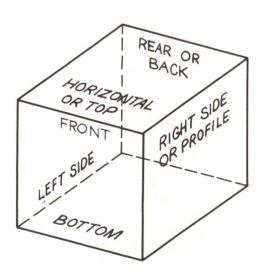

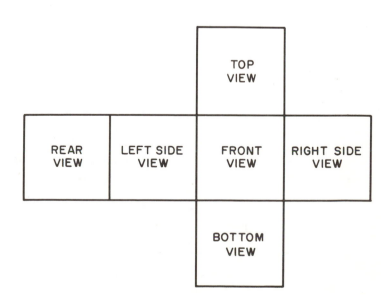

25

3.2 DEFINITION AND REPRESENTATION OF POINTS

A *point* is defined as a nondimensional location in space. It is physically represented by a small point, dot, or cross (+) and is identified by a single letter or number. A point can be located in the principal views by projecting it onto the principal planes of projection. Figure 3-2 illustrates three principal views of point A. The point A is labeled A_F in the front view, A_H in the top view, and A_P in the side view. The subscripts correspond respectively to the frontal, horizontal, and profile planes of projection.

This conventional method of projecting a point onto the principal planes of projection can be related to the familiar X, Y, Z coordinate system used in mathematics. Using the X, Y, Z coordinate system shown in Figure 3-2, the point A can be located with respect to the three axes.

- Height and width dimensions are always seen in the front view.
- Width and depth dimensions are always seen in the top view.
- Height and depth dimensions are always seen in the side view.

A constant relationship between views is that point A in the top view is always vertically aligned with point A in the front view and point A in the side view is always horizontally aligned with point A in the front view.

The projection lines between adjacent views are always perpendicular to the fold lines between the two adjacent planes of projection, as shown in Figure 3-2.

Figure 3-2 Principal Views of a Point

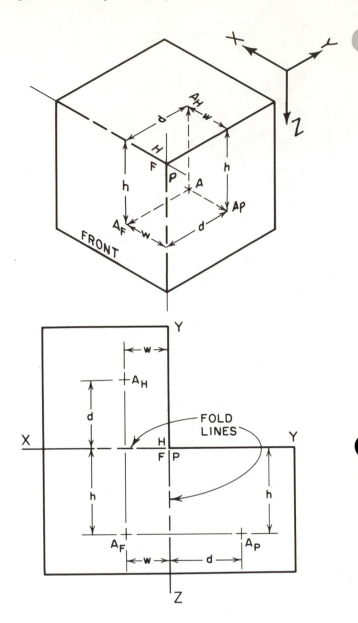

3.3 AUXILIARY VIEWS OF POINTS

If it is desired to locate a point as viewed from some direction other than one of the principal views, it is necessary to draw an auxiliary view. Any auxiliary plane that is perpendicular to one of the principal planes of projection but inclined to the other two principal planes is called a first or primary auxiliary plane. Any auxiliary plane that is perpendicular to a first auxiliary plane is called a second or successive auxiliary plane.

The projection planes of any two adjacent views are always perpendicular to each other, whether principal or auxiliary views.

In Figure 3-3, point A is seen orthographically projected onto an auxiliary plane that is perpendicular to the horizontal plane but inclined to the front and profile planes.

In this example, the first auxiliary view is projected from the top view. Once the view direction is established, the projection line from the principal view (top view, in this illustration) to the auxiliary view is drawn parallel to the view direction. The fold line H/1 is located at a convenient distance from the top view and is drawn perpendicular to the projection line.

The distance h is transferred from the front view to the auxiliary view since *the distance of point A below the horizontal plane is the same in both the front view and the first auxiliary view.*

26

Figure 3-3 First Auxiliary View of a Point

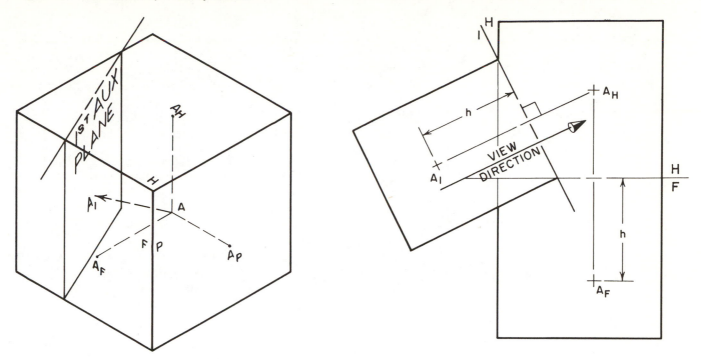

A second auxiliary plane is an oblique plane perpendicular to the first auxiliary plane and is inclined to all three principal planes of projection. Views projected on such oblique planes are second or successive auxiliaries. The relationships between second and first auxiliaries are the same as between first auxiliaries and principal views. Figure 3-4 illustrates a second auxiliary view of point A. Point A is the distance d behind the frontal plane. Point A is the distance t from the first auxiliary plane.

Figure 3-4 Second Auxiliary View of a Point

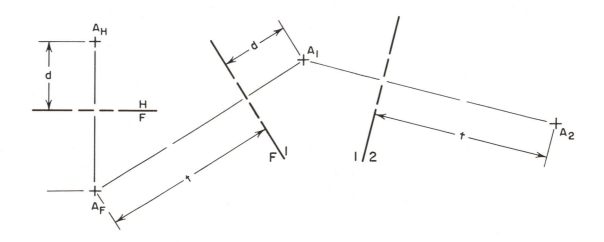

3.4 DEFINITION AND REPRESENTATION OF LINES

A *line* is the path of a point in space. It may be straight or curved. A straight line is determined by any two points. However, it is not limited by the two given points and is considered to extend infinitely beyond its given end points.

A minimum of two adjacent orthographic views is required to show specific information about a line. If the point or end view of a line, the true length of a line, or the slope of a line cannot be seen in any principal view, each can be found in an auxiliary view.

3.5 CLASSIFICATION OF LINES

A line can represent a center line, an axis, an intersection, the edge view of a surface, a surface limit, or any one of many graphical symbols. It is helpful to understand the properties of lines and their relationship to the planes of projection. *If a line is parallel to a plane of projection,*

it will appear true length (TL) when projected on that plane. See Figure 3-5.

- A line parallel to a frontal plane is called a *frontal line.* Such a line will appear true length in the front view.
- A line parallel to a horizontal plane is called a *horizontal line* and will appear true length in the top view.
- A line parallel to a profile plane is called a *profile line* and will appear true length in the side view.

An *inclined line* is parallel to one principal plane and inclined to the other two principal planes. Figure 3-5(b) shows a line parallel to the horizontal plane and inclined to the frontal and profile planes. The line is true length in the top view and foreshortened in the front and side views.

If a line is inclined to all three principal planes, it is called an *oblique line.* Figure 3-6 shows a line inclined to all three principal planes. It will appear foreshortened in any principal view.

Figure 3-5 Classification of Lines

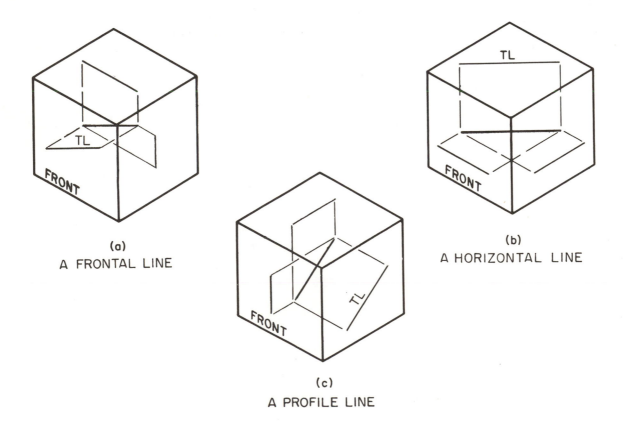

(a)
A FRONTAL LINE

(c)
A PROFILE LINE

(b)
A HORIZONTAL LINE

28

Figure 3-6 Principal Views of a Line

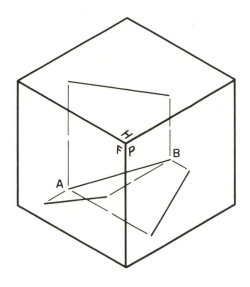

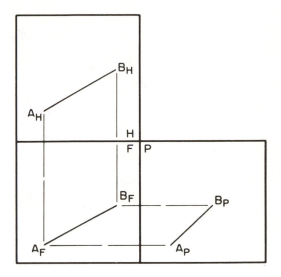

3.6 CHARACTERISTICS OF TWO LINES

When considering two lines together, several common relationships should be understood.

- Parallel lines [Figure 3-7(a)] are lines lying in the same plane and extending in the same direction. The distance between them is always the same.
- Intersecting lines [Figure 3-7(b)] are lines lying in the same plane with one point in common.

- Skew lines [Figure 3-7(c)] are nonparallel, nonintersecting lines.
- Perpendicular lines [Figure 3-7(d)] are intersecting or skew lines that have a 90° relationship to each other.

Two lines do not have to intersect to be perpendicular.

Figure 3-7 Characteristics of Two Lines

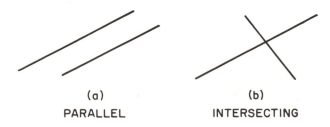

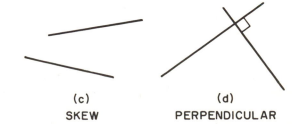

3.7 LOCATION OF A POINT ON A LINE

A point on a line in any given view has a definite relationship with the same point in adjacent views. This relationship is shown in Figure 3-8, in which there are three principal views and an auxiliary view of line *AB*. Point *P* is a specific point on the line. From the principles of orthographic projection, point *P* in the front view is horizontally aligned with point *P* in the profile view. Also, point *P* in the front view is vertically aligned with point *P* in the top view. From the rules of auxiliary projection, point *P* in the top view and point *P* in the auxiliary view will lie on a projection line that is parallel to all the projection lines between the top view and the auxiliary view.

3.8 INTERSECTING LINES

When two lines intersect, they have one point in common. Figure 3-9 shows three principal views and an auxiliary view of two intersecting lines *AB* and *CD*. The point of intersection in the front view must be horizontally aligned with the point of intersection in the profile view and must be vertically in line with the point of intersection in the top view. The projection line between the point of intersection in the profile view and the point of intersection in the auxiliary view must be parallel to the other projection lines between the two lines, as shown in Figure 3-9.

If the projection line joining the apparent points of intersection between two adjacent views is *not parallel* to the other projection lines, then the lines *do not* have one point in common and do not intersect (Figure 3-10).

Figure 3-8 Point on a Line

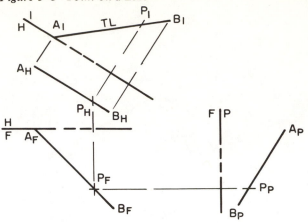

Figure 3-9 Intersecting Lines

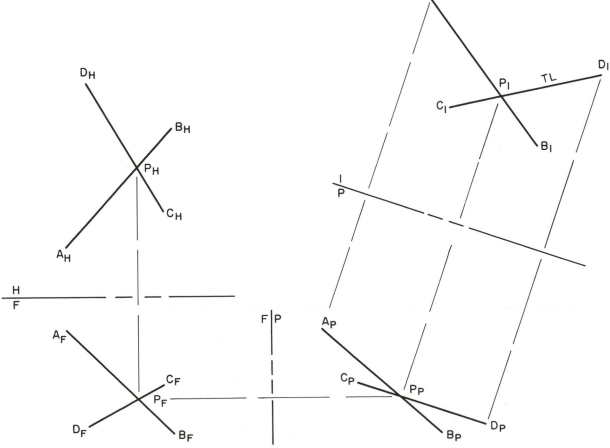

Figure 3-10   Nonintersecting Lines

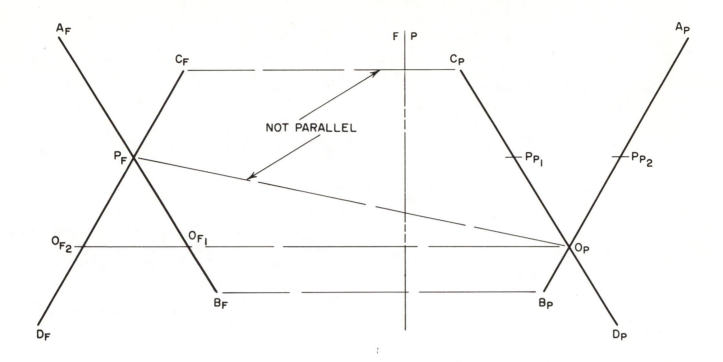

3.9 PARALLEL LINES

Parallel lines lie in the same plane but never intersect. *If lines are actually parallel in space they will appear parallel in any view*. See Figure 3-11. Note that the true length of one or both of the lines is not a requirement for determining whether or not the lines are parallel. Nonparallel lines can conceivably appear parallel in one view, but they will not appear parallel in all views.

In Figure 3-12, the two lines are parallel to the pro-

file plane and therefore *appear* parallel to each other in the front and top views. In the profile view, however, it is readily apparent that the two lines are not parallel.

In cases where lines are parallel to a principal plane and appear parallel to each other, it is very helpful to draw a true-length view of the lines to determine if they are parallel.

Figure 3-12   Nonparallel Lines Parallel to a Principal Plane

Figure 3-11   Parallel Lines

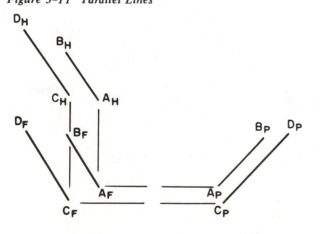

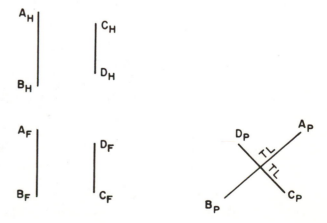

31

3.10 PERPENDICULAR LINES

When lines are perpendicular to each other, *they will appear perpendicular in any view in which one or both of the lines are true length.*

In Figure 3-13, line AB is true length in the top view. The angle between $A_H B_H$ and $C_H P_H$ is 90°. The lines are, therefore, perpendicular to each other. The point C could be rotated around line AB to any position and the angle between the lines will still be 90° and, of course, the lines are perpendicular.

To determine if two lines are perpendicular to each other, project the lines to a view in which one or both of the lines are true length. If the angle between the projected lines is 90°, the two lines are perpendicular. If the lines appear to be perpendicular, but neither line is true length, then the lines *are not* perpendicular.

3.11 TO CONSTRUCT THE TRUE-LENGTH VIEW AND THE POINT VIEW OF A LINE

If a line is perpendicular to a principal plane of projection, one principal view of it will be an end or point view and the adjacent view will show its true length.

If a line is parallel to one principal projection plane and is inclined to the other two, it will be true length in one principal view only.

If a line is oblique (inclined to all three projection planes), a true-length view can be obtained by a single auxiliary view, which is projected on a plane parallel to the line.

Figure 3-13   Perpendicular Lines

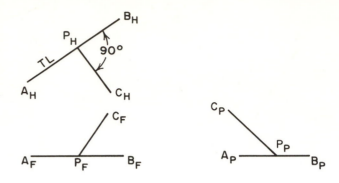

To obtain the point view of any line, it must be projected on a plane perpendicular to the line; that is, the view direction must be parallel to the true-length line. To find the point view of an oblique line, a second auxiliary view is necessary with the view direction parallel to the true length of the line as found in the first auxiliary view.

Example. (Figure 3-14): Given the oblique line AB. The first auxiliary view gives the true length of AB. A second auxiliary view taken in a direction parallel to the true-length view of the line gives the point view of the line.

Figure 3-14   Construction of True Length and Point View of a Line

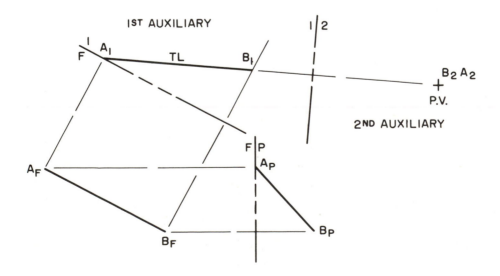

Figure 3-15 Definition of a Plane

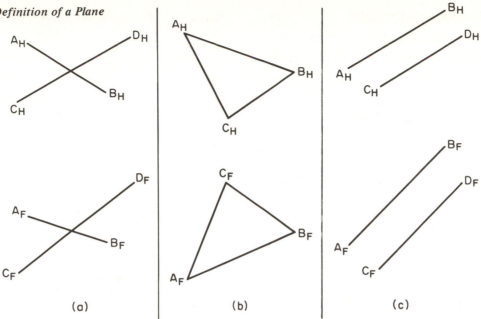

(a) (b) (c)

3.12 DEFINITION AND REPRESENTATION OF PLANES

In simple terms, a *plane* is a flat surface. Therefore, a straight line connecting any two points in the plane will lie wholly on the surface. There are three ways to define a plane.

- Two intersecting lines will define a plane [Figure 3-15(a)].
- Three points not on a straight line will define a plane [Figure 3-15(b)].
- Two parallel lines will define a plane [Figure 3-15(c)].

Usually, a plane surface is thought of as a finite surface such as the side of a box, but in the solution of spatial relationship problems, a plane is infinite, extending beyond the points or lines that define it.

3.13 CLASSIFICATION OF PLANES

Just as lines are classified by types, so are planes. A plane that is parallel to a frontal plane of projection is called a *frontal plane* and will appear true shape (*TS*) in the front view. A frontal plane is perpendicular to the horizontal and profile planes [Figure 3-16(a)].

A plane that is perpendicular to one principal plane but inclined to the other two is called an *inclined plane*. Figure 3-16(b) shows a plane perpendicular to the horizontal plane. It will appear in edge in the top view and will appear foreshortened in the front and side views.

A plane that is inclined to all three principal planes is called an *oblique plane*. An oblique plane will appear foreshortened in any principal view [Figure 3-16(c)].

Figure 3-16 Classification of Planes

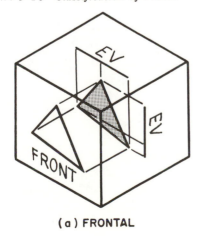

(a) FRONTAL   (b) INCLINED   (c) OBLIQUE

33

Figure 3-17   Locating a Line on a Plane   *Figure 3-18   Locating a Point on a Plane*

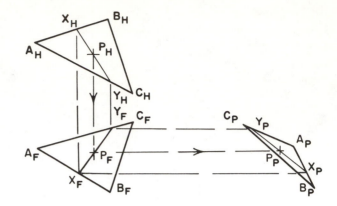

3.14   POINTS AND LINES ON PLANES

If a line on a plane is given in any view, it can be located in adjacent views. The line *ST*, shown on the front view of plane *ABC* in Figure 3-17, can be located in the adjacent views by projecting the end points of the line parallel to the projection lines between the views. The point *S* lies on line *AC* in the given front view. It must lie on line *AC* in the top and side views. Likewise, point *T* lies on line *BC* in the front view and must lie on line *BC* in the adjacent views.

If a point is given in any view of a plane, it can be located in adjacent views by first constructing a line through the point. In Figure 3-18, the point *P* is given on plane *ABC* in the top view. Any line *XY* on plane

ABC is drawn through point *P* and projected to the adjacent views as was done in Figure 3-17. The point *P* is then projected to the line *XY* from the given view to the adjacent views.

3.15   TO ESTABLISH A PLANE IN SPACE

A plane is determined by three points not in the same straight line, two parallel lines, or two intersecting lines. Thus, if any one of these geometric arrangements in space is represented on planes of projection, the location of a plane in space has been established. In Figure 3-19, the two intersecting lines, *AB* and *CD*, determine a plane.

Figure 3-19   Construction of a Plane in Space

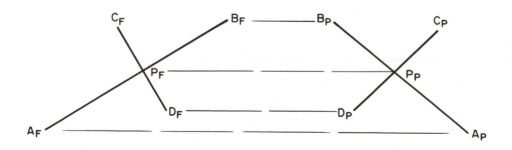

3.16 TO CONSTRUCT AN EDGE VIEW AND A TRUE-SHAPE VIEW OF A PLANE

A plane that is perpendicular to a principal plane will appear in edge when projected on that plane. To obtain an edge view of an oblique plane, it must be projected onto an auxiliary plane. The problem is to determine the view direction in order to obtain the edge view of the plane.

The important principle here is that by establishing a view direction parallel to a line of the plane that appears in true length in one of the principal views, the resulting auxiliary view will show the plane in edge. Pick up a triangle and look down an edge so that the edge appears as a point. Notice that the triangle appears in edge.

If none of the given lines of the plane appears true length in the principal views, a line can be constructed in the plane parallel to a principal plane. This constructed line will then appear true length in the adjacent principal view to which the line is parallel.

Example. (Figure 3-20): In Figure 3-20(a), there are two views of an oblique plane *ABC*. None of the given lines is true length in either view.

- Draw a horizontal line *CX* in plane *ABC* in the front view and project it to the top view [Figure 3-20 (b)].
- Line *CX* is in the plane *ABC* and is true length in the top view.
- To obtain the point view of line *CX*, project parallel to *CX* in the top view onto an auxiliary plane.
- This auxiliary view will show the point view of line *CX* and the edge view of the plane *ABC*.

In summary, to obtain the edge view of an oblique plane, first obtain the true-length view of a line in that plane and then project parallel to that true-length line.

To obtain the *true shape* of a plane, the line of sight is perpendicular to the plane. In other words, the lines of projection must be *perpendicular to the edge view* of the plane.

Figure 3-20   Construction of the Edge View of a Plane

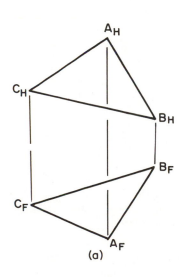

(a)

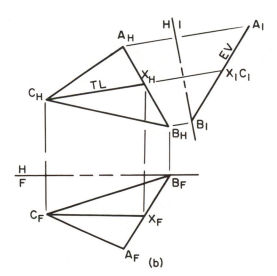

(b)

Figure 3-21   Construction of the True-Shape View of an Inclined Plane

An inclined plane will appear in edge in one of the principal views. Therefore, a single auxiliary view, projected in a direction perpendicular to the edge view, will show the true shape of the plane. This technique is illustrated in Figure 3-21. Plane *ABC* is shown in edge in the top view. By projecting from the edge view of plane *ABC* onto an auxiliary plane in a direction perpendicular to the edge view, the true-shape view is obtained.

To find the true shape of an oblique plane, two operations are required: first, obtain a view in which the oblique surface appears in edge; second, obtain a view in which the surface appears in true shape. These two operations are accomplished by drawing first and second auxiliary views.

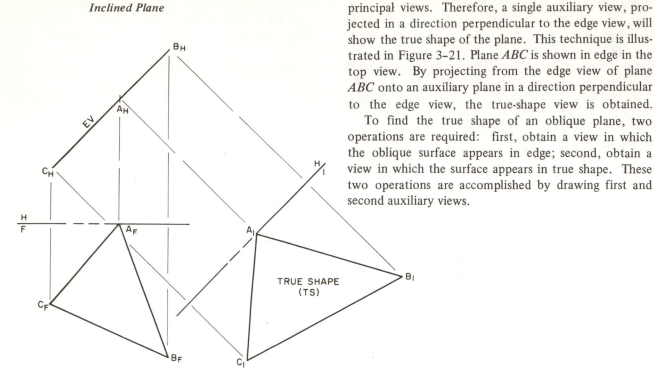

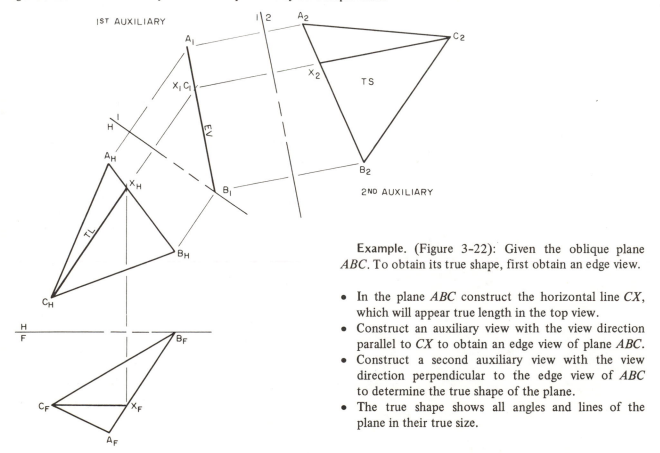

Figure 3-22   Construction of the True-Shape View of an Oblique Plane

Example. (Figure 3-22): Given the oblique plane *ABC*. To obtain its true shape, first obtain an edge view.

- In the plane *ABC* construct the horizontal line *CX*, which will appear true length in the top view.
- Construct an auxiliary view with the view direction parallel to *CX* to obtain an edge view of plane *ABC*.
- Construct a second auxiliary view with the view direction perpendicular to the edge view of *ABC* to determine the true shape of the plane.
- The true shape shows all angles and lines of the plane in their true size.

3.17 SOLIDS

The previous paragraphs on points, lines, and planes are all applicable to the representation of solids. Since the surfaces of all solid objects are a combination of points, lines, and planes, the same procedures for drawing principal and auxiliary views can be used. There are, however, some additional conventional practices that should be noted when making multiview drawings of solid objects.

3.18 SELECTION OF VIEWS

Careful consideration should be given to the selection of views to best describe an object. Usually a front, top, and right-side view will be adequate. Some objects can be completely described in one or two principal views. In Figure 3-23, the shape of a cylinder is completely described in the front and side views. A sphere can be described in one view.

Figure 3-23 Shape Description of a Cylinder

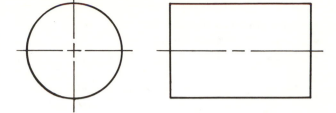

If inclined or oblique surfaces exist on the object, it may be necessary to add auxiliary views or partial views to the principal views to fully describe the object. In Figure 3-24, the three principal views are augmented by a partial auxiliary view showing the true shape of the inclined surface. Whatever views are used, they should be selected and oriented to best describe the shape of the object.

Figure 3-24 Complete Shape Description

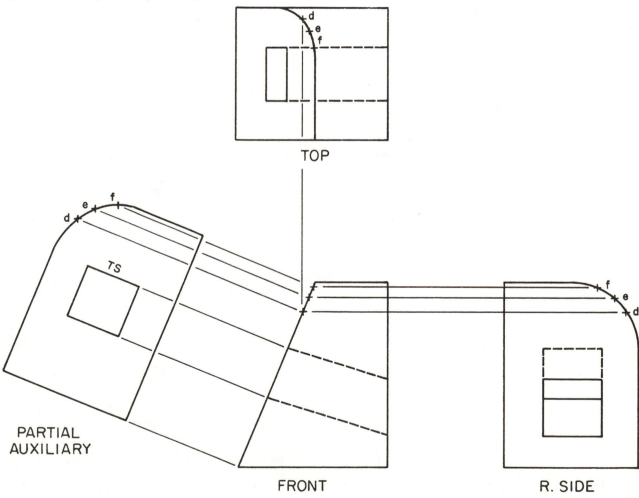

37

3.19 PRECEDENCE OF LINES

A line in a drawing of a solid object is caused by one of three conditions.

- The edge of a receding surface.
- The intersection of two surfaces.
- The surface limit of a curved surface.

Figure 3-25 illustrates these three conditions. Since every point of an object is seen in every view of an orthographic projection, some of the lines may be hidden on the drawing, but will be shown with the hidden-line symbol. The precedence of line symbols, if different lines coincide, is given as follows.

1. Object lines.
2. Hidden lines
3. Center lines and cutting-plane lines.
4. Break lines.
5. Dimension and extension lines.
6. Section lines.

3.20 ORDER OF DRAWING

For efficient preparation of a multiview orthographic drawing, the following step-by-step procedure is recommended.

1. Make a pictorial sketch of the object to help visualize the details.
2. Determine the combination of views needed to best describe the object.
3. Select a suitable scale and size of paper.
4. Block in the views with light outlines and center lines. Surfaces of revolution should have a center line shown. Every circle has crossing center lines to indicate its center.
5. Add the details of the object, building the views together. Avoid completing any one view alone.
6. Add lettering and titling. In making drawings of complex objects, it is helpful to identify each point as it is located.
7. Check the drawing.

Figure 3-26 illustrates the steps for making a multiview drawing of a solid.

Figure 3-25 Determination of Object Lines

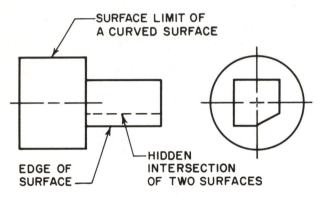

Figure 3-26 Order of Drawing

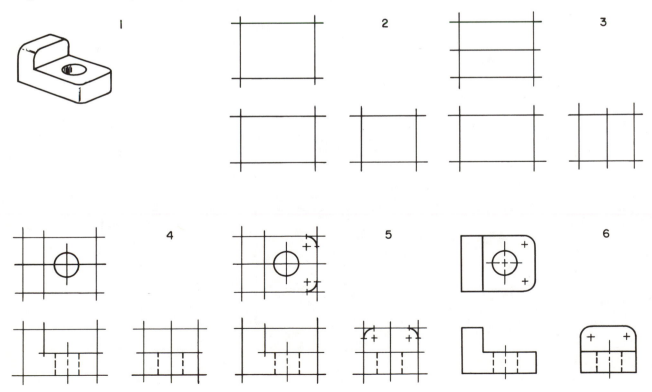

3.21 CONCLUSION

The fundamental spatial relationships of points, lines, planes, and solids have been systematically delineated in this chapter. The student should have a thorough understanding of orthographic projection, principal and auxiliary views, and a standard form of notation. The application of these fundamental spatial relationships to the solution of many practical engineering problems will be discussed in the following chapter.

Applied Spatial Relationships

4

The graphical solution of engineering problems is based on the fundamental spatial relationships of points, lines, and planes. By applying the principles of orthographic projection, many practical engineering problems involving spatial relationships can be solved.

4.1 TO CONSTRUCT A LINE PARALLEL TO A GIVEN LINE THROUGH A GIVEN POINT

If two lines are parallel in space, they will appear parallel in any view. Review Section 3.9.

Example. [Figure 4-1(a)]. Given the line *AB* and the point *C*, construct a line *CD* parallel to *AB*.

• Construct line *CD* parallel to *AB* in all views [Figure 4-1(b)].

Figure 4-1 Construction of Parallel Lines

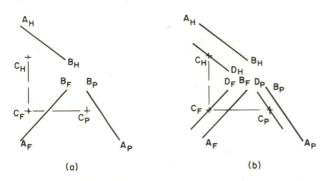

(a)   (b)

4.2 TO CONSTRUCT A LINE FROM A GIVEN POINT PERPENDICULAR TO A GIVEN LINE

This construction is based on the theorem that if two lines are perpendicular in space, *they must appear perpendicular in a view which shows the true length of one or both of the lines.* Figure 4-2(a) pictorially depicts this theorem. The construction is simplified if either the given or constructed lines appear true length in one of the principal views. If this condition does not exist, an auxiliary view is drawn to obtain the required true-length view of one of the lines. Review Section 3.10.

Example. [Figure 4-2(b)]: Draw a line from point *P* that will be perpendicular to the horizontal line *AB*.

• From point *P* in the top view, draw *PC* perpendicular to the true-length view of *AB*.

• C_F can be located anywhere on the projection line from C_H.

Figure 4-2 Construction of Perpendicular Lines

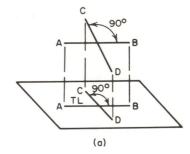

(a)

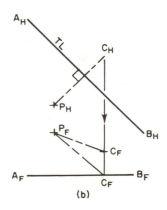

(b)

41

Figure 4–2 (continued)

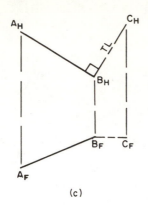

(c)

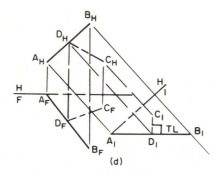

(d)

Example. [Figure 4–2(c)]: Draw a horizontal line *BC* perpendicular to the oblique line *AB*.

- Construct *BC* perpendicular to *AB* in the top view.
- In the front view construct the horizontal line *BC* so that *BC* is true length in the top view.

Example. [Figure 4–2(d)]: Draw a line *CD* from point *C* perpendicular to and intersecting the oblique line *AB*.

- Project a first auxiliary of line *AB* and point *C* so that *AB* is true length.
- Construct *CD* perpendicular to *AB* in this auxiliary.
- Project point *D* to the given views.

4.3 TO CONSTRUCT A LINE PARALLEL TO A GIVEN PLANE AND CONVERSELY TO CONSTRUCT A PLANE PARALLEL TO A GIVEN LINE OR LINES

If a line is constructed parallel to any line in a given plane, the constructed line will also be parallel to the given plane.

Conversely, if a plane contains a line parallel to a given external line, the plane is parallel to the given line.

Example. [Figure 4–3(a)]: Given the lines *AB* and *MN*, construct a plane containing *MN* and parallel to *AB*.

- Construct *PQ* parallel to *AB* through any point *O* on *MN*.
- The intersecting lines *MN* and *PQ* determine a plane that is parallel to *AB*.

Example. [Figure 4–3(b)]: Given the lines *AB* and *CD* and the point *P*, construct a plane containing point *P* that will be parallel to lines *AB* and *CD*.

- Construct lines *UV* and *XY* through *P* and parallel to *AB* and *CD*, respectively.
- The intersecting lines *UV* and *XY* determine a plane that is parallel to the given lines *AB* and *CD*.
- Note that *AB* and *CD* in this cases are skew lines.

Figure 4–3 *Construction of Parallel Line and Plane*

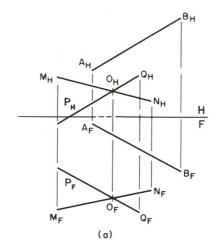

(a)

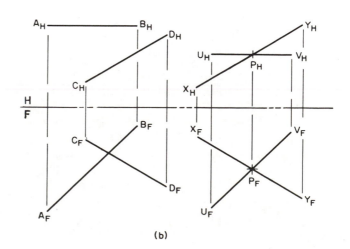

(b)

Figure 4-4 Construction of a Line Perpendicular to a Plane

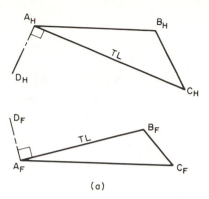

(a)

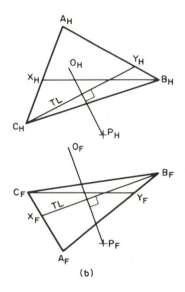

(b)

4.4 TO CONSTRUCT A LINE PERPENDICULAR TO A PLANE

A line perpendicular to a plane is perpendicular to every line in that plane. Consequently, if in adjacent views a line is drawn perpendicular to the true-length views of two intersecting lines lying in the plane, a line perpendicular to the plane is established.

Example. [Figure 4-4(a)]: Given the plane *ABC*, construct a line perpendicular to plane *ABC*. In this case, where the true length of a line exists in both given views, the solution is simple.

- Construct *DA* perpendicular to the true length of the horizontal line *AC* in the top view and to the true length of the frontal line *AB* in the front view.
- The line *DA* is perpendicular to the plane *ABC*.

Example. [Figure 4-4(b)]: Construct a line from point *P* perpendicular to the given plane *ABC*.

- Construct in both views the horizontal line *CY* and the frontal line *BX* in the plane *ABC*.
- Construct line *PO* perpendicular to the true length of the horizontal line *CY* in the top view.
- Construct line *PO* perpendicular to the true length of the frontal line *BX* in the front view.
- These two lines are the top and front views, respectively, of the perpendicular to the plane *ABC* from the point *P*.

4.5 TO CONSTRUCT A PLANE THROUGH A POINT AND PERPENDICULAR TO A LINE

Since a line perpendicular to a plane is perpendicular to every line in that plane, if two lines are drawn through a given point, both perpendicular to the given line, they will determine the required plane.

Example. (Figure 4-5): Construct a plane containing point *P* and perpendicular to line *AB*.

- Construct horizontal line *PO* perpendicular to *AB* in the top view.
- Construct frontal line *PN* perpendicular to *AB* in the front view.
- Lines *PO* and *PN* determine the required plane. Refer to Figure 4-2 for the construction of a line from a given point perpendicular to a given line.

Figure 4-5 Construction of a Plane Perpendicular to a Line

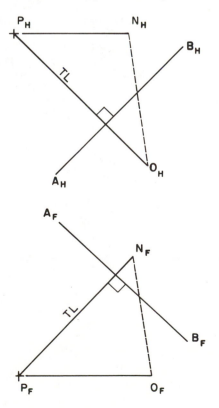

4.6 TO DETERMINE THE POINT AT WHICH A LINE PIERCES A PLANE AND THE VISIBILITY OF THE LINE

One method of determining the point where a line pierces a plane is to construct a view showing the plane in edge and note where the line pierces the plane.

Example. (Figure 4–6): Given two views of line AB and a plane CDE.

- Construct an auxiliary view showing the edge view of plane CDE.
- Determine where line AB pierces plane CDE. Line AB intersects the edge view of the plane at piercing point P.
- The point P can be projected to the principal views.
- Visibility of the line and the plane may be determined by the technique illustrated in Figure 4–8.

Another method used to locate piercing points is called the *cutting-plane method*. Figure 4–7 shows a pictorial drawing of a plane CDE and a line AB. To obtain the point where line AB pierces plane CDE, pass a vertical plane $MNOP$, containing line AB, cutting the original plane CDE. Plane $MNOP$, which contains line

Figure 4–6   Piercing Point of a Line and a Plane—Edge View Method

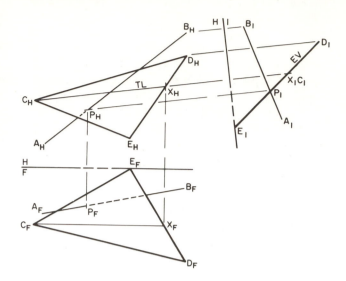

AB, intersects plane CDE along the line XY. The piercing point P is the point at which line AB intersects line XY. P is the only point common to plane CDE and line AB.

Figure 4–7   Pictorial Drawing of Line Piercing a Plane—Cutting Plane Method

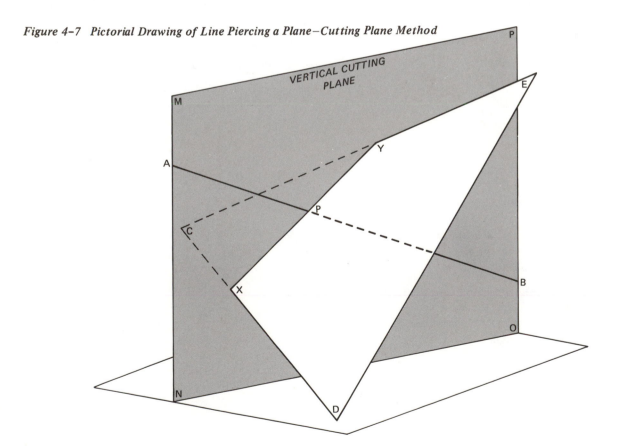

Example. (Figure 4-8): Given two views of the plane *CDE* and the end points of the line, *A* and *B* [Figure 4-8(a)].

- A vertical cutting plane containing the line *AB* is passed through plane *CDE* and cuts *CDE* along the line *XY*. Notice that line *XY* coincides with line *AB* in the top view and also notice that the line *XY* can be located in the front view by projection.
- Line *AB* will pierce the plane *CDE* somewhere along line *XY*. This point cannot be located in the top view because the lines coincide.
- The piercing point *P* can be located in the front view by drawing line *AB* and noting the point where it intersects line *XY* [Figure 4-8(b)]. This point of intersection is the piercing point.

The visibility of crossing (nonintersecting) lines can be determined by examining the point of crossing in any two adjacent views. In Figure 4-8, it is apparent that line *AB* and line *CD* are not intersecting lines. However, in the top view, the two lines do cross. Since they do not intersect, one of the lines must be closer to the viewer's eye or be visible at the point of crossing.

To determine which line is visible, project the point of crossing from the top view to the front view. This projection line will intersect line *AB* in the front view before it reaches line *CD*. *This indicates that at the point of crossing in the top view, line AB is higher or above line CD and is therefore visible.*

It should be emphasized that the visibility of crossing lines in any given view cannot be determined from that given view. It is necessary to examine an adjacent view as explained above.

To determine the visibility of line *AB* and the plane *CDE* [Figure 4-8(b)], the procedure described above is extended to each crossing point in each view.

Line *AB* has already been determined to be visible above line *CD* in the top view. Thus line *AB* must be visible from point *A* to the piercing point *P*.

Since the line pierces the plane, it must become hidden from point *P* until it emerges from under the plane at the point where it crosses under line *CE*. This establishes the visibility of line *AB* and plane *CDE* in the top view.

The visibility in the front view must be determined by reversing the procedure and projecting from a point of crossing in the front view to the top view.

Figure 4-8   Piercing Point of a Line and a Plane—Cutting Plane Method

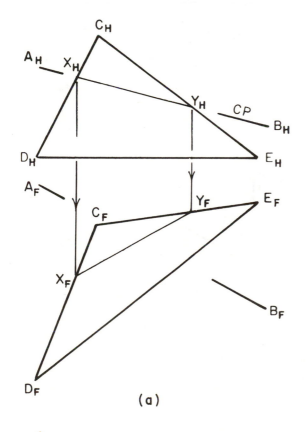

(a)

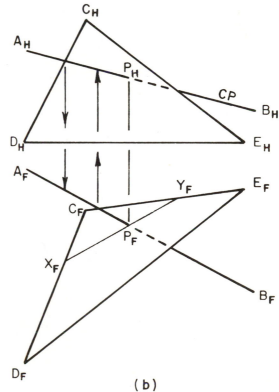

(b)

45

4.7 TO DETERMINE THE INTERSECTION OF TWO PLANES

Any two points common to both planes will determine a straight line that lies in both planes and is the intersection of the two planes. A point common to the two planes is determined by finding the point in which a line of one plane pierces the second plane (Figure 4-8).

Example. (Figure 4-9): Given the planes *ABC* and *LMNO*.

- Determine the piercing point *P* of the line *LM* with the plane *ABC* using the method of Figure 4-8; likewise, determine the piercing point *S* for the line *NO*.
- Project points *P* and *S* to the top view. Line *PS* is the required intersection.
- Determine visibility and show hidden lines.

Figure 4-9 To Determine the Intersection of Two Planes

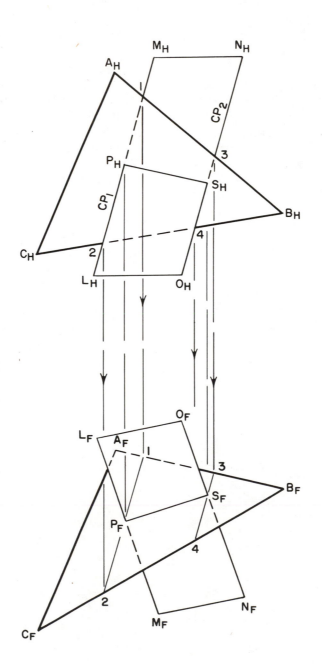

46

4.8 ROTATION

Rotation is another graphical approach to solving many spatial problems. When using auxiliary views, the viewer changes position and the object remains stationary; whereas when using rotation, the viewer remains stationary and the object is rotated about an axis of rotation to the desired position.

The following principles provide the basis for solving any problem by use of rotation.

1. An axis of rotation must be determined before an object can be rotated.
2. The point view of the axis of rotation must be determined.
3. The projection of the object on the plane showing the point view of the axis of rotation changes position during rotation but does not change its original size or shape.
4. The dimensions of the object parallel to any true-length projection of the axis of rotation remain constant during rotation of the object about the axis.

The concept of rotation is best understood by first rotating a point about an axis. In Figure 4-10(a), point Q has been rotated 90° clockwise about axis AX. It is obvious that as a point rotates about an axis, it remains in a plane perpendicular to the axis.

Then, in Figure 4-10(a), the point Q can be rotated about the axis AX in the front view since AX appears as a point.

In the top view, the point Q moves but remains in a plane perpendicular to the axis AX. Since the axis is true length in the top view, the plane perpendicular to the axis will appear in edge as a line perpendicular to AX.

Point Q rotated 90° clockwise to its new position would then appear at Q'_F as shown in the front view and as Q'_H projected to the top view.

If the axis of rotation is not given true length, it is necessary to construct auxiliary views so that the axis appears as a point [Figure 4-10(b)]. Rotation can then be used to find the new position of the point Q by rotating the point about the point view of the axis as in the previous example. The point Q in its rotated position can then be projected back to the given views.

Figure 4-10 Rotation of a Point About an Axis

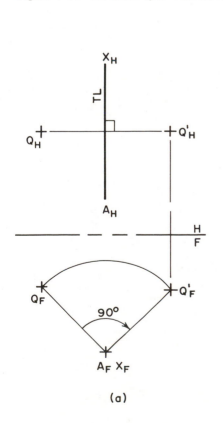

(a)

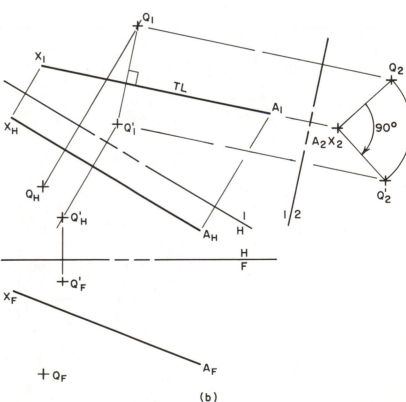

(b)

47

The rotation of a line to find its true length is illustrated in Figure 4-11. This is probably the most widely used application of rotation in fundamental spatial relationships.

The line is rotated until it is parallel to a principal plane and is then projected on that principal plane to obtain a true-length view of the line. In the pictorial illustration, visualize a right triangle *ABX* with the hypotenuse *AB*, which is an oblique line. *AB* will not appear true length in any principal view. If the entire triangle is rotated about its vertical axis, *AX*, until the triangle is parallel to the frontal plane of projection, the line *AB* will appear true length in the front view. Notice that when the triangle rotates, point *A* does *not* move and point *B* remains in the same horizontal plane.

Figure 4-11 Rotation of a Line About an Axis

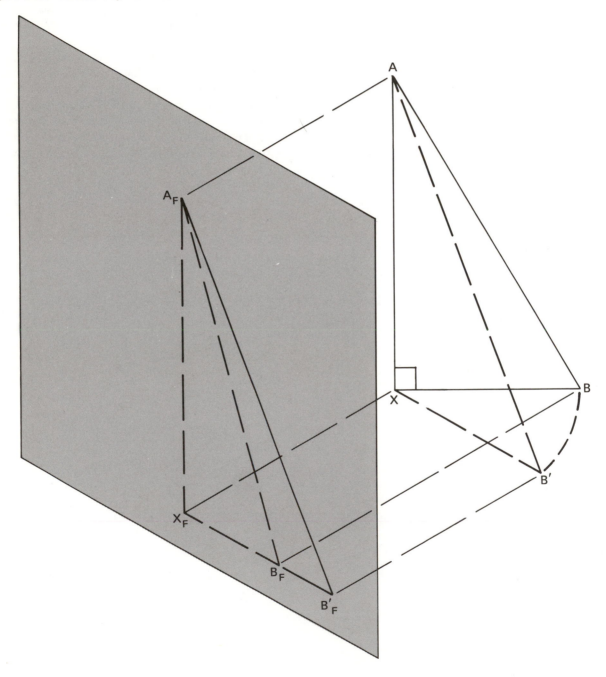

Figure 4-12(a) shows the front and top view of an oblique line *AB*. To find the true length by rotation, rotate line *AB* about a *vertical* axis through point *A* until *AB* is parallel to the *frontal* plane.

Note that in the top view, point *A* remains fixed and point *B* rotates about the axis until *AB* is parallel to the frontal plane.

In the front view, point *A* also remains fixed, and as point *B* rotates, it remains in the same *horizontal* plane. Thus, in the front view, *B* moves horizontally until it is vertically beneath the new position of *B* in the top view.

Since line *AB'* is now parallel to the frontal plane, it appears true length in the front view.

The same principle applies to rotating a line in any two adjacent views. Figure 4-12(b) shows an oblique line *AB* in the front and profile views. Rotate line *AB* about a *horizontal* axis through point *A* until line *AB* is parallel to the *frontal* plane. In the front view, *A* does not move and *B* moves vertically until it is horizontally in line with *B* in the side view. The resulting front view of *AB'* is true length of the line.

Figure 4-12 Rotation of a Line

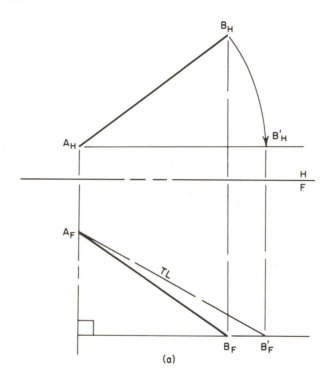

(a)

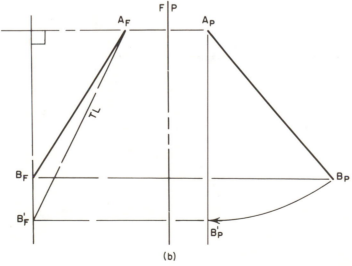

(b)

The true shape of a plane may also be found by rotation (Figure 4–13).

- A plane will appear true shape when rotated into or parallel to a plane of projection.
- To find the true shape of the oblique plane ABC, first construct an auxiliary view showing the plane in edge.
- Draw a horizontal line DA in the plane ABC. The line DA will be true length in the top view.
- The edge view of plane ABC will then appear in the auxiliary plane perpendicular to line DA.

- In the edge view, rotate the points B and C about the axis DA until the two points are in a horizontal plane.
- Project the rotated positions of B and C back to the top view until they intersect lines drawn through the top-view positions of B and C and perpendicular to the axis DA.

The two points have been rotated about the axis AD as was illustrated in Figure 4–10. This rotation results in the true shape of the given plane, $A_H B'_H C'_H$.

Figure 4–13 *True Shape of a Plane by Rotation*

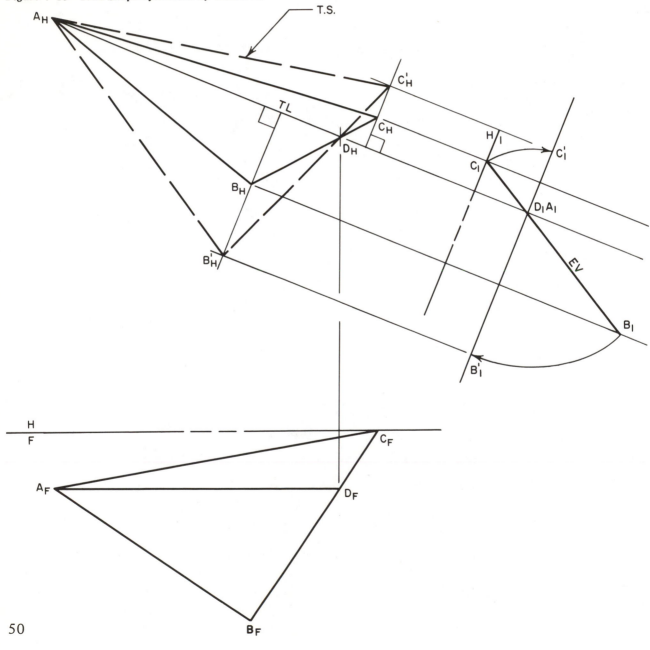

50

The true shape of a plane can also be found by double rotation as illustrated in Figure 4-14. This method requires no auxiliary views but is more involved graphically.

The plane is first rotated so that a true-length line in the plane appears as a point in an adjacent view. In Figure 4-14, the plane is rotated about a horizontal profile axis in the front view until the true-length line CX is vertical and appears as a point in the top view. The plane in this first rotated position will appear in edge in the top view.

The plane can then be rotated a second time by rotating it about a vertical axis in the top view until the edge view $A'_H C_H B'_H$ is parallel to the front plane ($A'_H C''_H B''_H$). The plane $A'_F B''_F C''_F$ is the true shape of the plane ABC.

Figure 4-14   True Shape of a Plane by Double Rotation

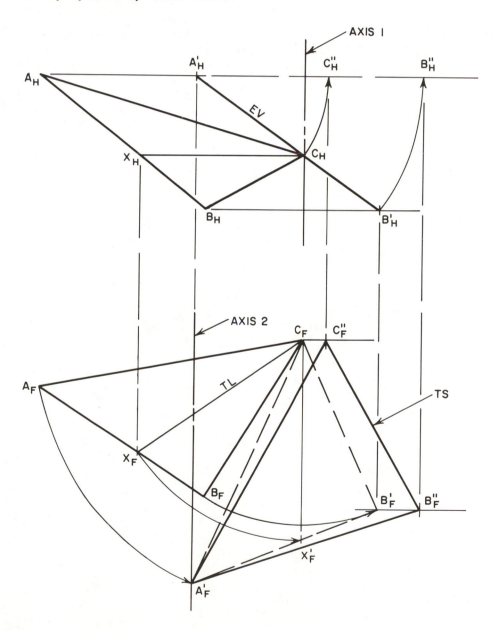

4.9 DISTANCE FROM A POINT TO A LINE

The shortest distance from a point to a line is the perpendicular distance from the point to the line.

- Construct a view with the given line in its true length (first auxiliary view in Figure 4–15).
- Construct a perpendicular P_1X_1 to the line. This line represents the shortest distance between the point and the line.

The true length of this distance can be found in either of two ways.

1. By drawing an auxiliary view showing the true length of the line *PX*.
2. By rotation.

In Figure 4–15, the line *PX* is rotated until it is parallel to the frontal plane and appears true length in the front view.

Figure 4–15 Distance from a Point to a Line—Line Method

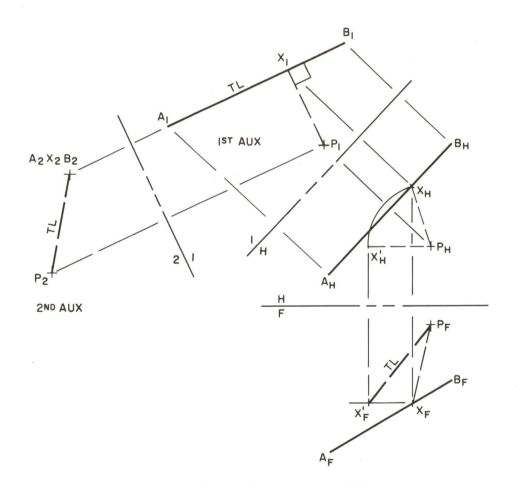

Another method of finding the distance between a point and a line is to find the true-shape view of the plane containing both the point and the line. Figure 4-16 shows line *AB* and point *P* forming a plane *ABP*.

The first auxiliary shows plane *ABP* in edge, and the second auxiliary is a true-shape view of the plane. In this second auxiliary view, the perpendicular distance from *P* to line *AB* is shown true length.

Figure 4-16 Distance from a Point to a Line—Plane Method

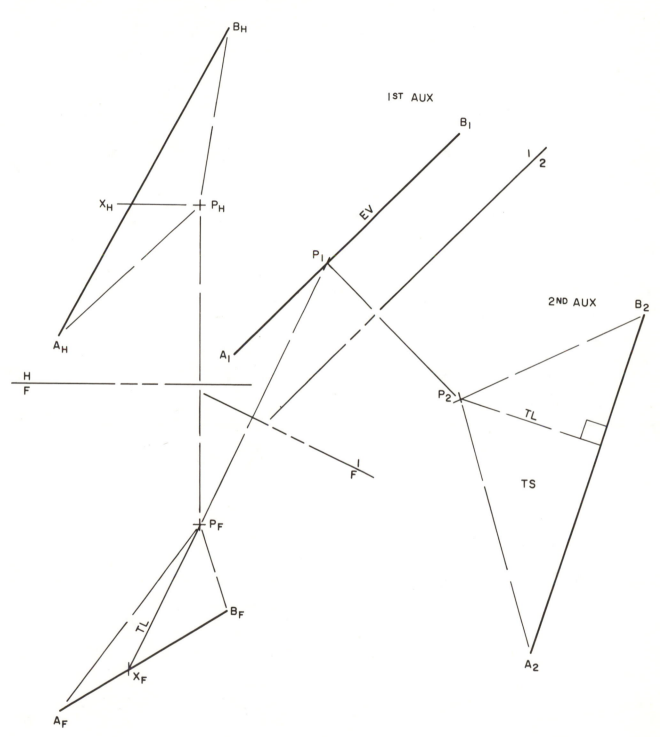

53

4.10 ANGLE BETWEEN TWO INTERSECTING LINES

To determine the angle between two intersecting lines it is necessary only to obtain the true-shape view of the plane containing those lines. Figure 4-17 illustrates this problem. The true angle between the two lines can be measured in the second auxiliary view, which shows the true length of lines *AB* and *CD*.

Figure 4-17   Angle Between Two Intersecting Lines

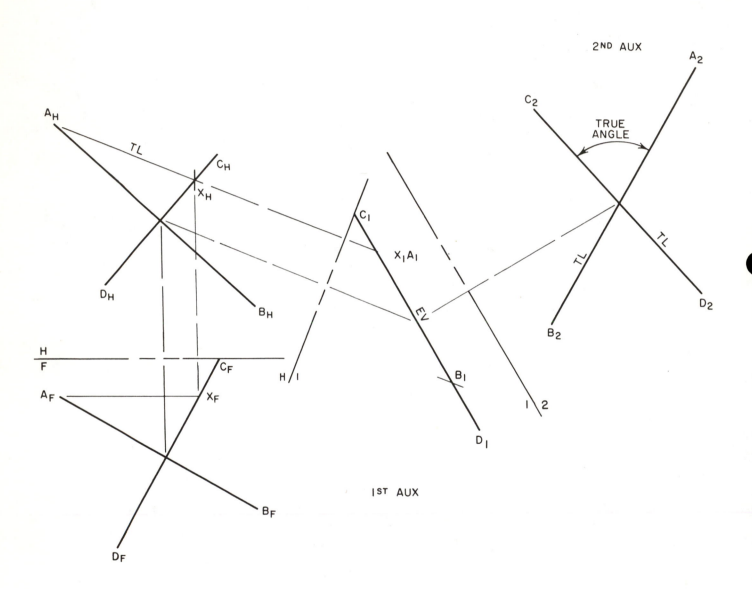

4.11 ANGLE BETWEEN A LINE AND A PRINCIPAL PLANE

To find the angle between a line and a principal plane, _it is necessary to obtain a view that shows the true length of the given line and an edge view of the principal plane._ The horizontal plane appears in edge in any view projected from the top view.

To find the angle between a line and a horizontal plane, construct an auxiliary view projected from the top view showing the true length of the given line. The required angle is the angle between the true length of the line and the edge view of the horizontal plane (Figure 4-18).

The frontal plane appears in edge in any view pro-

jected from the front view. To find the angle between a line and the frontal plane, construct an auxiliary view projected from the front view showing the true length of the given line. The required angle is the angle between the true length of the line and the edge view of the frontal plane.

The profile plane appears in edge in any view projected from the side view. To find the angle between a line and the profile plane, construct an auxiliary view projected from the side view showing the true length of the given line. The required angle is the angle between the true length of the line and the edge view of the profile plane.

Figure 4-18 Angle Between a Line and a Horizontal Plane

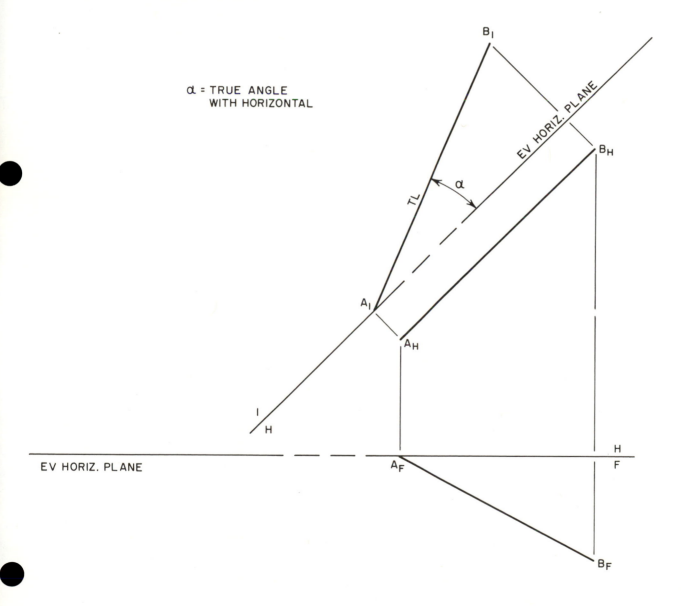

55

It is also possible to find the angle that a line makes with a principal plane by rotating the line. In Figure 4-19, line *AB* is rotated about a vertical axis until it is parallel to the frontal plane. In the front view, the horizontal plane is in edge and the line *AB'* is true length. The true angle between line *AB* and the horizontal plane is the angle between the edge view of the horizontal plane and the true-length line *AB'*.

Figure 4-19   Angle Between a Line and a Horizontal Plane—Rotation Method

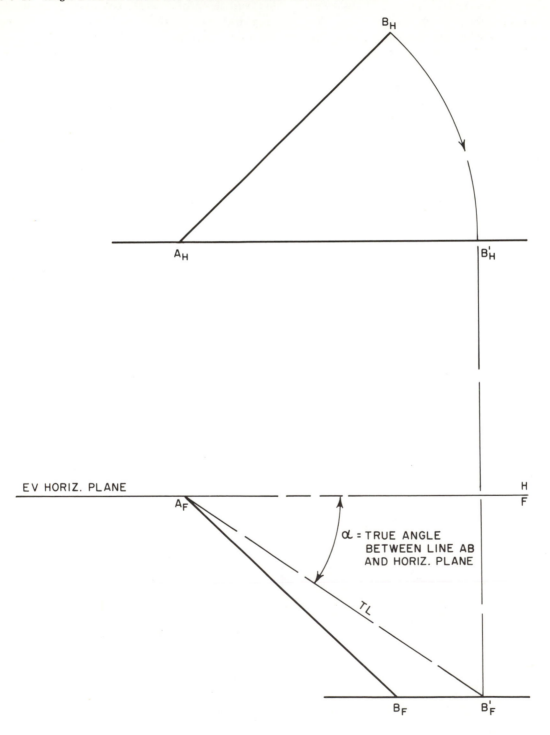

56

4.12 SLOPE, SLOPE ANGLE, GRADE, AND BEARING OF A LINE

A frequently used practical application of the previous problem is to find the slope of a line. *The slope of a line is defined as the deflection of a line from the horizontal.* Slope is expressed as a fraction, rise over run, which is the tangent of the slope angle. Slope angle is expressed in degrees, and grade is the slope expressed as a percentage. A positive or negative sign indicates whether the line is sloping up or down.

In Figure 4-20, line *AB* is an oblique line. To find the slope of *AB*, a first auxiliary view is drawn showing the edge view of the horizontal plane and the true length of the line *AB*. The slope is the fraction rise/run (50/100), as shown, and is expressed as the fraction 1/2. By inspection in the front view, the line *AB* slopes up. Therefore,

the slope is +1/2. The slope angle, α, is measured as +27°, and the grade is +50%.

The bearing of a line is defined as the map direction of the line measured from a reference, usually a north-south line. Since the slope of the line has no influence on the bearing of the line, the angle between the given line and the reference line is always measured in the top view.

The bearing of a line is conventionally the acute angle measured from north or south. In Figure 4-20, the bearing of the line *AB* is N 60° W. It should be noted that the bearing of line *BA* is in the opposite direction or S 60° E. Bearings in the four cardinal directions are expressed as due north, due east, due south, or due west.

Figure 4-20   Slope and Bearing of a Line

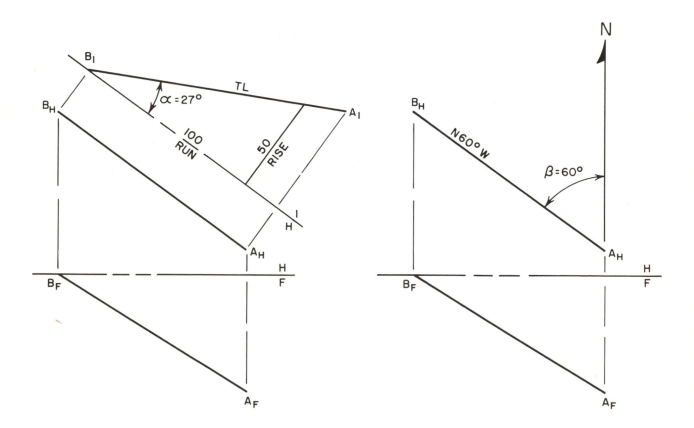

4.13 ANGLE BETWEEN A LINE AND ANY PLANE

To find the angle between a line and any plane, it is necessary to obtain a view that shows the true length of the line and an edge view of the plane. The required angle can then be measured.

The general solution to this problem requires three auxiliary views; however, either of two approaches may be used. See Figures 4-21 and 4-22.

In Figure 4-21, to find the angle between line DE and plane ABC, construct first and second auxiliary views to obtain the true-shape view of the plane.

- Any view projected from the second auxiliary view will show the plane in edge.
- Project to a third auxiliary view showing the true length of the given line. This view will show the true length of the line and an edge view of the plane.
- The angle between the line and the plane can be measured as shown.

Figure 4-21   Angle Between a Line and Any Plane—First Method

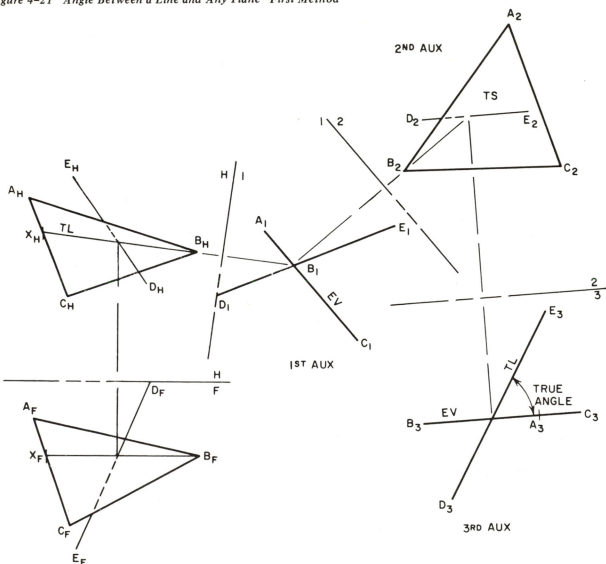

In Figure 4–22, to find the angle between line *DE* and plane *ABC*, construct first and second auxiliary views to obtain the point view of the line.

- Any view projected from the second auxiliary will show the line true length.

- Project to a third auxiliary view so that the plane will appear in edge. This view will show the true length of the line and an edge view of the plane.
- The angle between the line and the plane can be measured as shown.

Figure 4–22 Angle Between a Line and Any Plane–Second Method

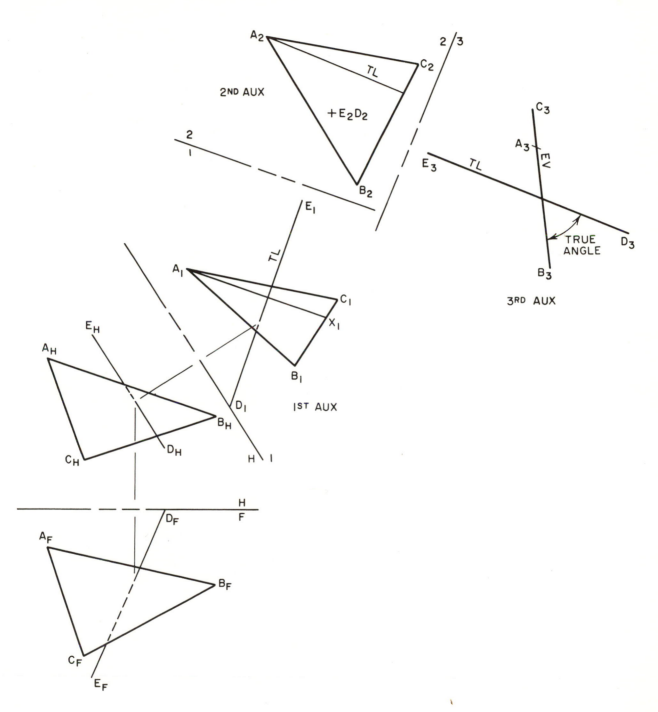

4.14 DISTANCE FROM A POINT TO A PLANE

To find the distance from a point to a plane, it is only necessary to construct a view showing the plane in edge.

The required distance is the perpendicular distance from the point to the edge view of the plane. Figure 4-23 illustrates this problem.

Figure 4-23   Distance from a Point to a Plane

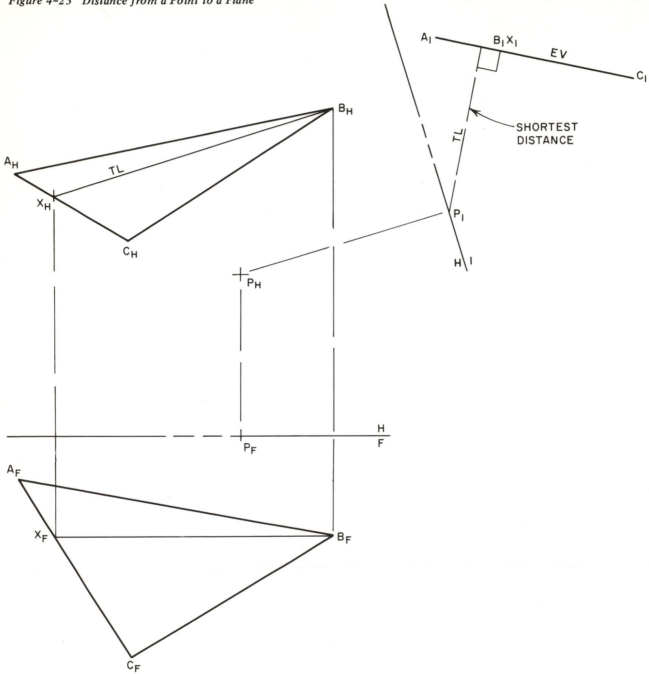

4.15 ANGLE BETWEEN A PLANE AND A PRINCIPAL PLANE

The angle between any plane and a principal plane can be observed in the view in which both planes appear in edge. The horizontal plane appears in edge in any auxiliary view projected from the top view. Likewise, the frontal plane appears in edge in any auxiliary projected from the front view, and the profile plane appears in edge in any auxiliary projected from the side view.

To find the angle a given plane makes with the horizontal plane, construct a first auxiliary view, projected from the top view, showing the plane in edge. The angle between the given plane and the horizontal plane can be measured in this first auxiliary view. Figure 4-24 illustrates this method of finding the angle between a given plane *ABC* and a horizontal plane.

The angle a plane makes with the frontal plane can be measured in an auxiliary projected from the front view, and the angle a plane makes with the profile plane can be measured in an auxiliary projected from the side view.

Figure 4-24   Angle Between a Plane and a Horizontal Plane

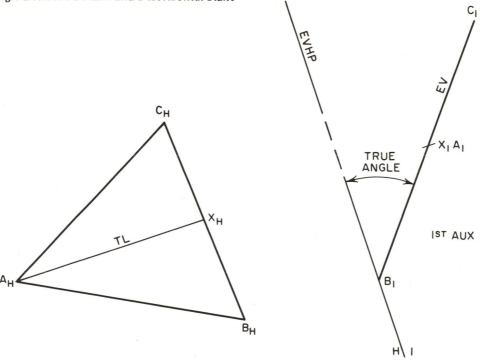

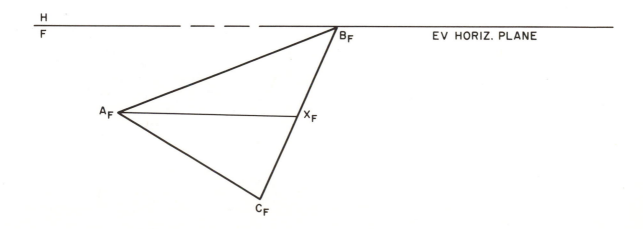

4.16 STRIKE AND DIP OF A PLANE

A practical problem encountered in general engineering practice is to determine the strike and dip of a plane. These terms refer to the direction and slope of a plane and define the orientation of the plane with respect to a north–south reference and the horizontal.

Strike is defined as the bearing of a horizontal line in the plane. It is conventionally measured as the acute angle from north.

Dip is the angle the given plane makes with the horizontal plane with the general compass direction of the down slope included, such as NE or SW. This general dip direction is measured perpendicular to a horizontal line of the plane in the top view toward the low side of the plane.

Figure 4-25 shows the front and top views of plane ABC. To find the strike and dip of the plane in Figure 4-25, proceed as in Figure 4-24 to find the angle between any plane and the horizontal plane.

- The strike of plane *ABC* is the bearing of the horizontal line *AX*, which is N 71° E.
- The dip angle of the plane is the true angle between the edge view of the plane and the horizontal plane in the first auxiliary view, which is 40°.
- The dip direction of the plane is found in the top view by constructing the perpendicular to the strike line toward the low side of the plane, which is NW.

The strike of the plane is N 71° E, and the dip of the plane is 40° NW.

Figure 4-25 Strike and Dip of a Plane

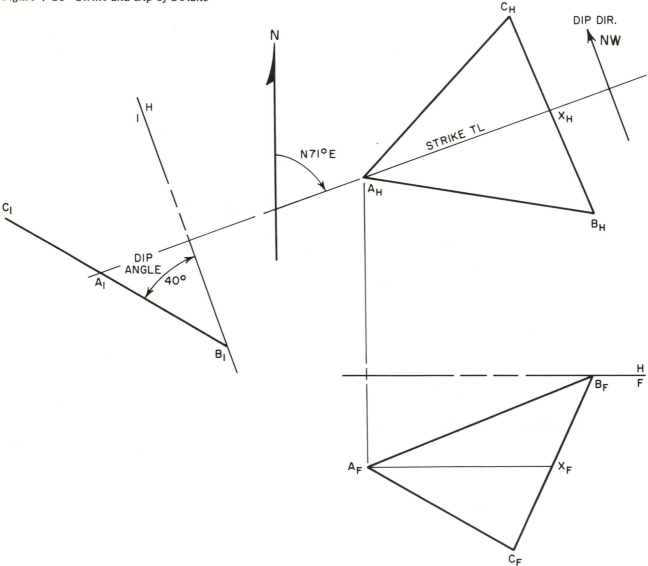

4.17 ANGLE BETWEEN TWO PLANES

To find the angle between two planes, it is necessary to find a view showing both planes in edge. It is possible to view both planes in edge by obtaining a point view of their line of intersection.

Figure 4–26 shows two planes, *ABC* and *BCD*. The first auxiliary view shows the true length of the line of intersection *BC*. The second auxiliary view shows the point view of line *BC*. Both planes will appear in edge in this second auxiliary view, and the angle between the two planes can be measured as shown.

Figure 4-26 Angle Between Any Two Planes

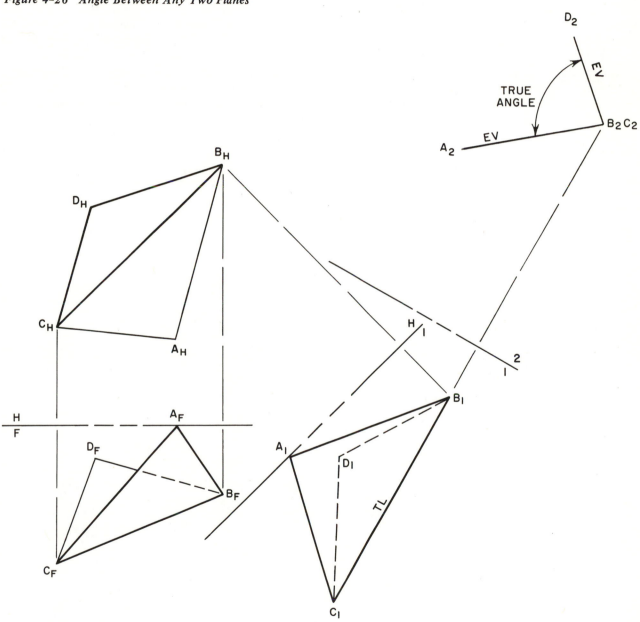

63

4.18 SHORTEST DISTANCE BETWEEN TWO SKEW LINES

One of the more important applications of the principles of perpendicularity if found in the solution of problems involving shortest connectors. The solution of these problems allows us to determine graphically rhe clearance or interference between lines in space representing shafts, control cables, pipes, or any other arrangement of lines. The advantage of a graphical solution to these problems is readily apparent when compared to analytical solutions.

In Figure 4-27, two nonparallel, nonintersecting, or skew lines, AB and CD, are shown in the front and top views. It is required to find the shortest distance between them or the shortest line connecting them.

The shortest distance between two skew lines will be a line that is perpendicular to both skew lines.

One method of finding the shortest distance between two skew lines is to construct a view showing the point view of one of the lines. The shortest distance is the perpendicular distance from the point view of the one line to the other line.

The lines do not appear in their true length in either of the principal views.

- Obtain the true length of line AB in a first auxiliary view.
- Project to a second auxiliary for the point view of the line.
- The shortest distance from line AB to line CD is Y_2X_2, the perpendicular from the point at A_2B_2 to the line C_2D_2.

The shortest connector can be located in the principal views by projecting back through the first auxiliary view to the principal views. Point Y cannot be projected back to the first auxiliary because it is not definitely located on line AB in the second auxiliary. However, in the first auxiliary view, line AB is true length. In the first auxiliary view, line XY is perpendicular to both AB and CD since the shortest intersecting line is perpendicular to both given lines. Line AB is true length in the first auxiliary view and Y_1 can be located on A_1B_1 by constructing the perpendicular from X_1 to A_1B_1. Point Y can then be projected back to the principal views.

Figure 4-27   Distance Between Two Skew Lines—Point View Method

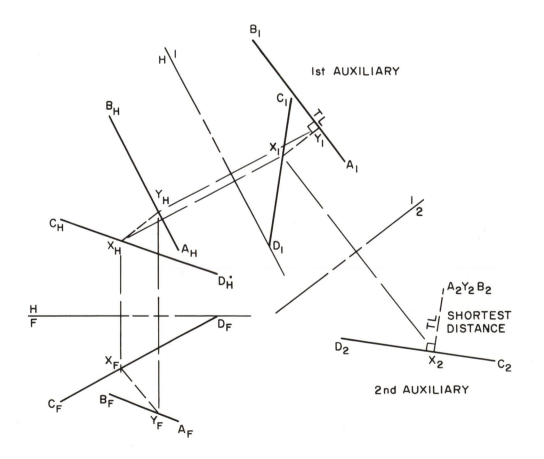

Another method of finding the shortest distance between two skew lines is to construct a plane that contains one of the lines and is parallel to the other line, and then find the edge view of the plane.

The distance between the edge view of the plane and the other skew line will be the shortest distance between the two skew lines.

Figure 4-28 shows the two skew lines *AB* and *CD*. To find the shortest distance between two lines by the parallel method:

- *DE* is drawn parallel to *AB*. *CD* and *DE* define a plane containing *CD* and parallel to *AB*.
- A first auxiliary view is drawn showing plane *CDE* in edge. The shortest *distance* between the plane and line *AB* is seen in this view.

If it is desired to find the location of the shortest *connector* or line between the two skew lines, draw a second auxiliary view obtaining a true-shape view of plane *CDE*. The point view of the shortest connector, Y_2X_2, can be projected back to the first auxiliary view and to the principal views.

Figure 4-28   Distance Between Two Skew Lines–Parallel Plane Method

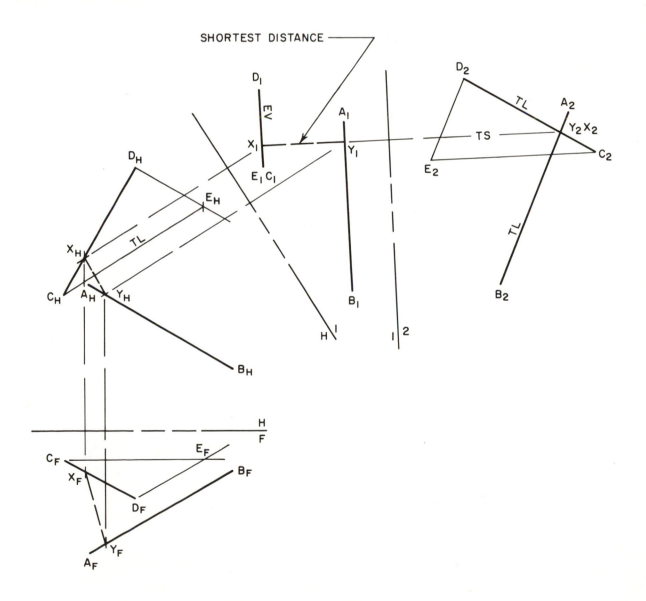

65

4.19 SHORTEST HORIZONTAL DISTANCE BETWEEN TWO SKEW LINES

The shortest horizontal distance between two skew lines can be found by the same general method as illustrated in Figure 4-28. However, since the shortest *horizontal* distance is required in this case, *it is necessary to project* *the second auxiliary view parallel to the horizontal as seen in Figure 4-29.*

The edge view of the horizontal plane is seen in the first auxiliary projected from the top view.

The point where the two lines cross in the second auxiliary view is the point view of the shortest horizontal line and can be projected back to the given views.

Figure 4-29 Shortest Horizontal Distance Between Two Skew Lines

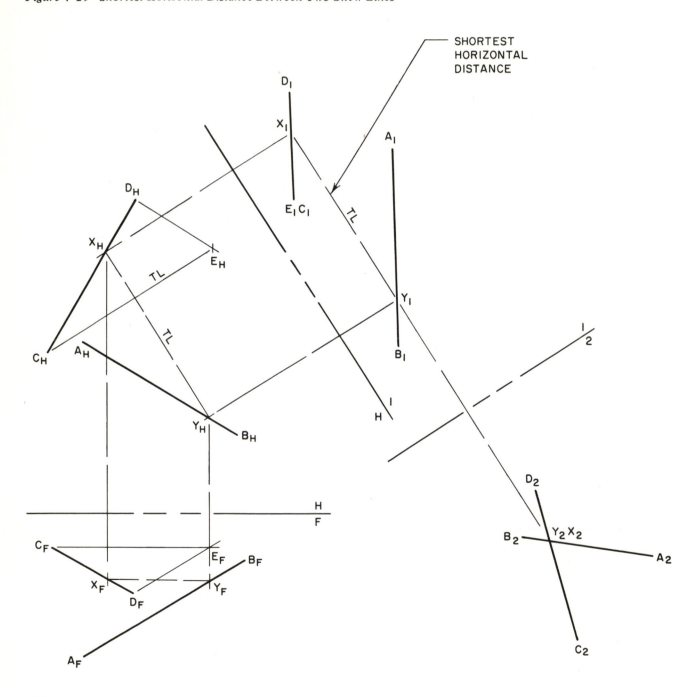

4.20 SHORTEST GRADE DISTANCE BETWEEN TWO SKEW LINES

To determine the shortest line at any specified grade between two lines, it is necessary to use the parallel plane method.

In Figure 4-30, to find the shortest 30% grade line between skew lines *AB* and *CD*.

- Project from the *top view* so that the horizontal plane will appear in edge and line *AB* will be parallel to a plane containing the other line.
- Construct a 30% grade line with respect to the horizontal plane in the first auxiliary view. Since there are two possibilities, construct this line in the

direction that most nearly approaches the perpendicular to the two given lines.

- The shortest grade *distance* will appear true length in the first auxiliary view.
- To locate the position of the shortest 30% grade line *XY*, project a second auxiliary parallel to the direction of the grade line.
- Project to the other views the points *X* and *Y* from the point of crossing of the two lines in the second auxiliary view.

Care must be taken if a positive or negative grade is specified, since the shortest negative grade would be the other 30% grade connector and may not be the shortest 30% grade line.

Figure 4-30 Shortest Grade Distance Between Two Skew Lines

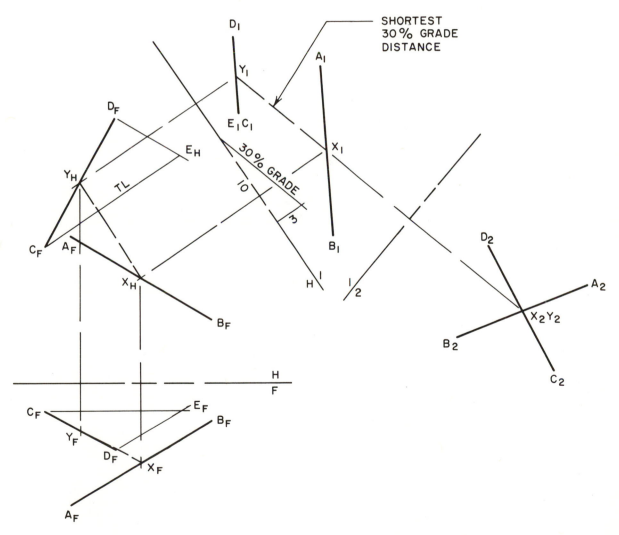

67

Intersections and Developments

5

Methods for finding piercing points and plane intersections were discussed in Chapter 3. By extending the use of these basic constructions, it is possible to find the line of intersection between intersecting geometric shapes.

5.1 INTERSECTIONS OF PLANES AND PRISMS

In Figure 5-1(a), the line of intersection in the front view coincides with the edge view (*EV*) of plane *ABCD*. In the top view, the piercing points of the vertical edges of the prism are represented by points 1, 2, and 3, and the line of intersection is the triangle 1-2-3. The side view is completed by projecting points 1, 2, and 3 and determining the visibility as shown in Figure 5-1(b).

Figure 5-1 Intersection of a Plane and a Prism—Edge View Method

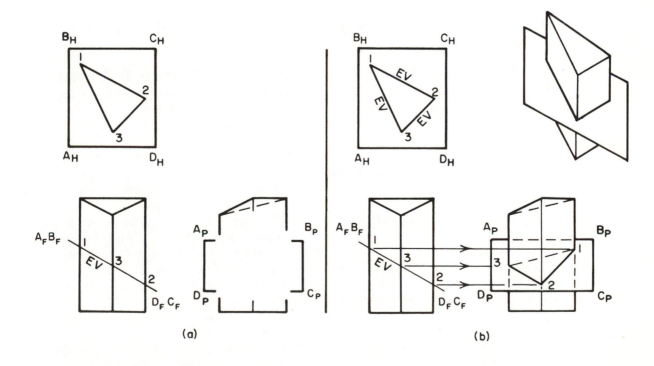

(a)   (b)

69

Figure 5-2 Intersection of a Plane and a Prism—Cutting Plane Method

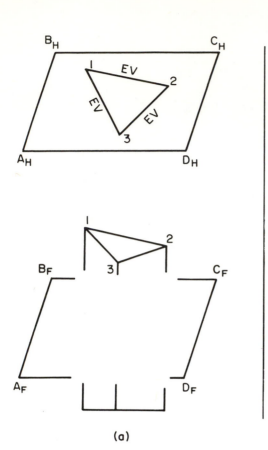

(a)

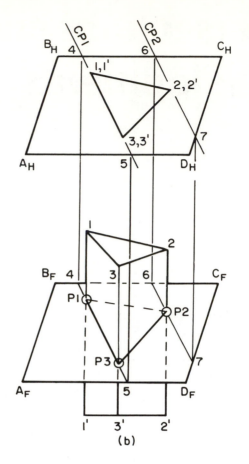

(b)

The cutting-plane method is used to determine the line of intersection between the plane and prism in Figure 5-2.

- Piercing point *P*1 and *P*3 are found by passing a vertical cutting plane *CP*1 through the edges 1-1′ and 3-3′ of the prism as shown in Figure 5-2(b).
- Piercing point *P*2 is found by passing a vertical cutting plane *CP*2 through the edge 2-2′ of the prism.
- The front view is completed by connecting the piercing points and determining the visibility.

In Figure 5-3, the plane and prism are both in oblique positions.

- An auxiliary view is constructed to determine the *EV* of the oblique plane.
- The piercing points *P*1, *P*2, and *P*3, where the edges 1-1′, 2-2′, and 3-3′ of the oblique prism pierce the oblique plane, are obvious.
- The distances *D*1, *D*2, and *D*3 can then be transferred to find *P*1, *P*2, and *P*3 in the top view.
- Project *P*1, *P*2, and *P*3 to the front view.

Determine the visibility and complete the views.

5.2 INTERSECTION OF TWO PRISMS

The intersection of a square prism and a triangular prism is illustrated in Figure 5-4. By inspection it can be seen that the top view is complete as shown in Figure 5-4(a).

- Edges 1-1′, 2-2′, and 4-4′ of the triangular prism pierce surfaces of the square prism at the piercing points *P*1, *P*2, and *P*4, respectively, and may be projected to the front view.
- The edge 3-5 of the square prism pierces the triangular prism. Using the edge view method, it is necessary to construct a partial auxiliary view showing the right section of the triangular prism as illustrated in Figure 5-4(b) to determine piercing points *P*3 and *P*5.
- Project *P*3 and *P*5 from the auxiliary view to the front view.
- Connect the piercing points in sequential order to determine the lines of intersection of the prisms, determine visibility, and complete the drawing.

Figure 5-3 Intersection of an Oblique Plane and Prism

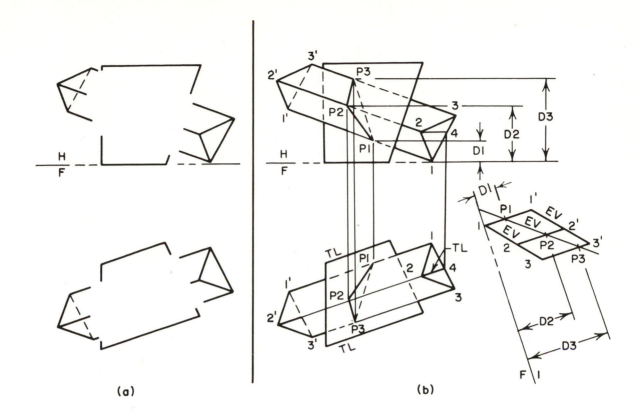

(a)

(b)

Figure 5-4 Intersection of Two Prisms

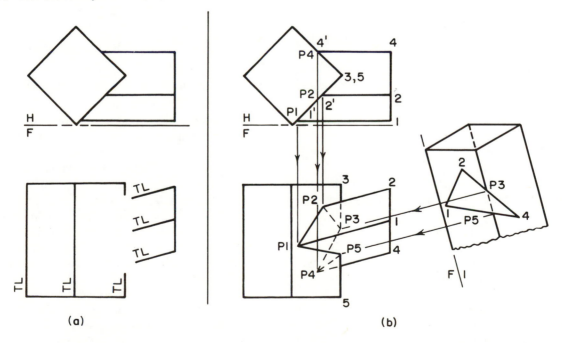

(a)

(b)

71

Figure 5-5   Intersection of a Plane and Cylinder

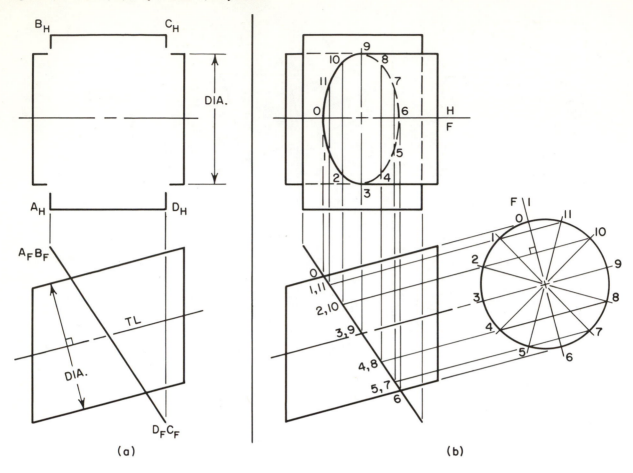

(a)

(b)

5.3   INTERSECTION OF A PLANE AND CYLINDER

The following rules may be applied to plane and right circular cylinder intersections.

If a cylinder is intersected by

1. A plane parallel to the axis, the resulting intersecting is two parallel straight lines.
2. A plane perpendicular to the axis, the intersection is a circle.
3. A plane inclined to the axis, the intersection is an ellipse.

In Figure 5-5, the plane *ABCD* that intersects the cylinder is inclined to the axis and appears on edge in the front view. The curve of intersection coincides with the straight line 0-6 in the front view.

• An auxiliary view is constructed to obtain a right section of the cylinder, which is then divided into twelve equal parts as shown.
• Project points from the auxiliary view to the front view.
• Project these points from the front view to the top view.

• Transfer corresponding distances from the auxiliary view to the top view to plot the elliptical curve of intersection.

5.4   INTERSECTION OF A PRISM AND CYLINDER

In Figure 5-6, the lateral surfaces of the triangular prism appear on edge in the top view, and the lines 1-5, 5-9, and 9-1 are the lines of intersection. Vertical cutting planes parallel to the axis of the cylinder cut intersecting straight line elements from the prism and cylinder to establish points on the curves of intersection. Elements from the prism that intersect elements from the cylinder on the curve of intersection can be projected directly to the front view. However, to obtain corresponding elements for the cylinder, the auxiliary view must be constructed.

Cutting planes, necessary to determine the limit of the intersection as well as the limit of visibility, must be carefully selected. Cutting planes 1, 5, and 9 are necessary to determine the limits. Cutting plane 5 also determines the limits of visibility.

72

Figure 5-6 Intersection of a Prism and Cylinder

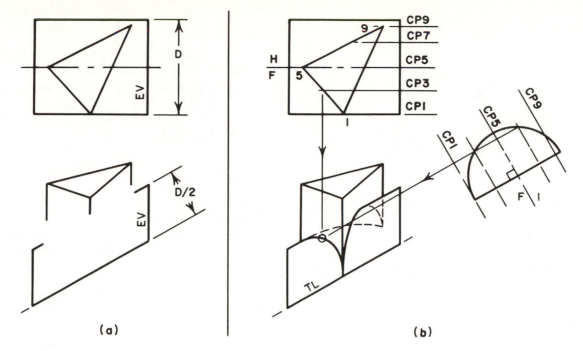

(a)   (b)

5.5 INTERSECTION OF TWO CYLINDERS

Two cylinders with axes perpendicular to each other are shown in Figure 5-7. In the top view, the curve of intersection is the circle representing the small cylinder.

- Cutting planes parallel to the axes of both cylinders are then located as shown in Figure 5-7(b) so as to cut straight line intersecting elements for locating points on the curve of intersection.

- An end view of the large cylinder is constructed, and the cutting planes are transferred to it as shown in the illustration.
- Elements from the small cylinder are projected from the top view.
- Corresponding elements from the large cylinder are projected from the end view to locate the necessary points on the curve of intersection.

Figure 5-7 Intersection of Two Cylinders

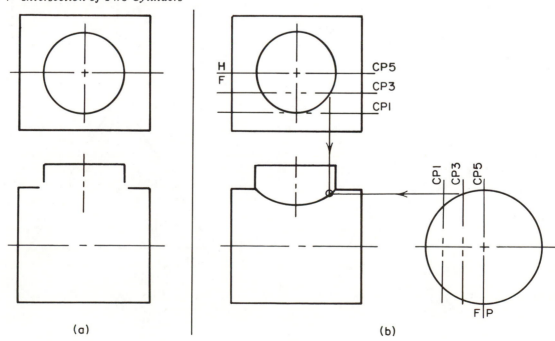

(a)   (b)

73

Figure 5-8 Intersection of Two Cylinders

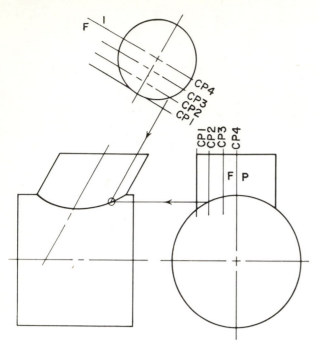

Two cylinders whose axes are not perpendicular are shown in Figure 5-8. The curve of intersection is found by the same procedure as in the previous illustration.

5.6 INTERSECTION OF A PLANE AND A RIGHT CIRCULAR CONE

A right circular cone and an intersecting plane are shown in Figure 5-9(a). If a plane cuts *all* elements of the cone but is not perpendicular to the cone axis, the curve of intersection is an ellipse.

- This curve may be determined by passing cutting planes perpendicular to the axis of the cone or through the vertex of the cone.
- In Figure 5-9(b), the front view shows the edges of cutting planes through the vertex. Each cutting plane cuts a straight line from the plane and from the cone.
- The intersection of these two lines determines a point on the curve of intersection. This is illustrated by point 4′ on the cone element *V*-4 in Figure 5-9(b).
- Locate sufficient cone elements in the front and top views to provide enough points delineating the curve of intersection.

Figure 5-9 Intersection of a Plane and Cone

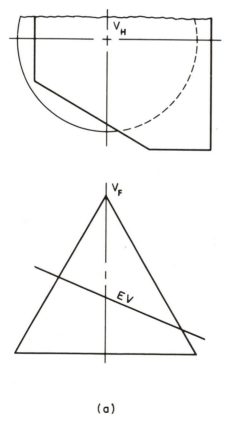

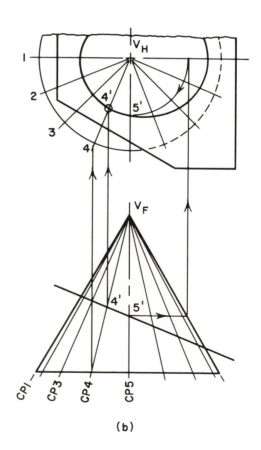

(a)

(b)

74

Figure 5-10 Intersection of a Prism and Cone

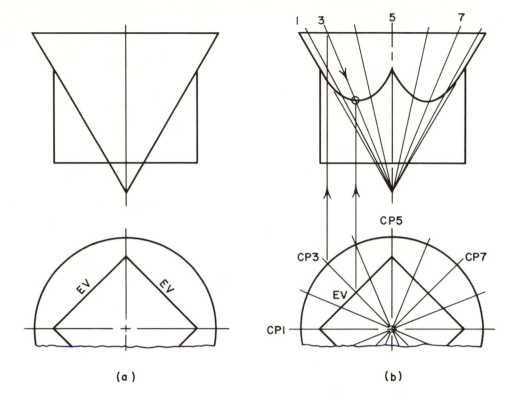

(a)

(b)

- The location of point $5'$ on the curve can be located in the top view by counter-rotation as shown.

5.7 INTERSECTION OF A PRISM AND CONE

The intersection of a prism and cone is complete in the partial front view of Figure 5-10. To determine the intersection in the horizontal view, the solution is similar to that of the preceding illustration. The cutting planes appear on edge in the front view in order to cut straight line elements from the cone and the prism. Cutting planes 1, 3, 5, and 7 establish critical points on the line of intersection. Note that cutting planes 3 and 7 locate the points on the curve of intersection closest to the vertex. This intersection can also be found by passing cutting planes perpendicular to the axis of the cone.

5.8 INTERSECTION OF TWO CONES

Figure 5-11(a) illustrates how a vertical cutting plane is passed through vertices $V1$ and $V2$ of the cones to find the upper limiting point on the curve of intersection. In

Figure 5-11(b), horizontal cutting planes are located as shown in the front view to cut circles from each of the cones. The radii are obtained from the front view as shown. Each pair of circles establishes two points on the curve of intersection.

5.9 INTERSECTION OF A CONE AND CYLINDER

Vertical cutting planes located as shown in Figure 5-12(b) will cut circles from the cone and straight lines from the cylinder. The radii for drawing the circles are obtained from the top view. A partial side view of the cylinder is drawn to provide a right section, which is used for locating the straight line elements of the cylinder. The straight line elements of the cylinder are then projected to the corresponding circles of the cone in the front view to locate points on the curve of intersection.

After a point on the curve of intersection is located in the front view, the corresponding point in the top view can be found by projecting back to the appropriate cutting plane as indicated in the illustration.

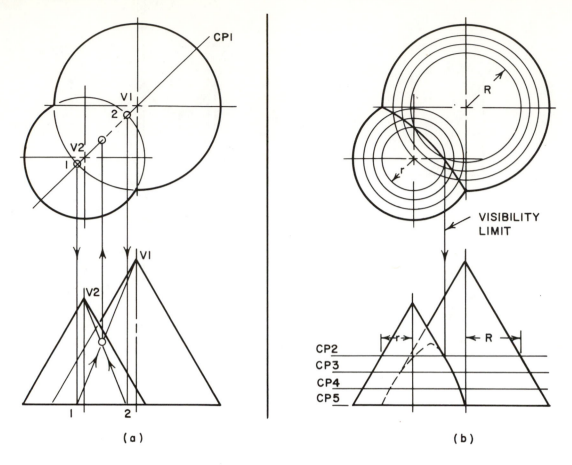

Figure 5-11   Intersection of Two Cones

(a)

(b)

Figure 5-12   Intersection of a Cone and Cylinder

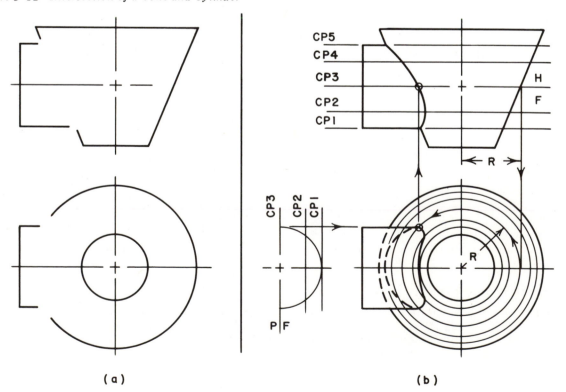

(a)

(b)

76

Figure 5-13 *Intersection of a Cylinder and Sphere*

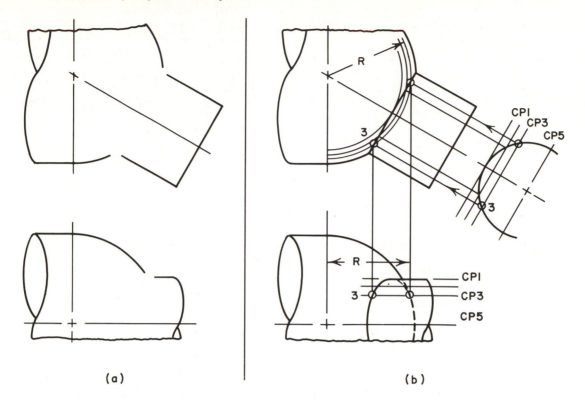

(a)   (b)

5.10 INTERSECTION OF A CYLINDER AND SPHERE

Figure 5-13 illustrates the intersection of a cylinder with a sphere. Horizontal cutting planes located as shown in the front view cut circles from the sphere and straight lines from the cylinder. The circles in the top view of the sphere are drawn with radii obtained from the front view as illustrated. It is necessary to draw a partial right section of the cylinder in order to locate the straight line elements of the cylinder. The intersections of corresponding circles and straight line elements produce points on the curve of intersection in the top view. These points are then projected back to the corresponding cutting-plane lines to locate the curve of intersection in the front view.

5.11 DEVELOPMENTS

In the manufacturing of component parts of a device or structure, it is frequently necessary to make a development or pattern of the part to determine its true size and shape. Only developable geometric shapes such as prisms, cylinders, pyramids, cones, and transition pieces will be considered in the remainder of this chapter.

5.12 DEVELOPMENT OF A RIGHT PRISM

The right square prism from Figure 5-4 has been redrawn as Figure 5-14. The edge view of a right section is indi-

is indicated in the front view, and the true shape of the right section is shown in the top view.

- A line coincident with the edge view of the right section is extended to the right. This line is called the stretch-out line.
- The true-length distances between the lateral edges 1-1′, 2-2′, 3-3′, and 4-4′ are transferred from the right section to the stretch-out line.
- The true lengths of the lateral edges are projected from the front view to the development.
- Points $P1$, $P2$, $P3$, $P4$, and $P5$ are projected as indicated. Points $P3$ and $P5$ are on the lateral edge 3-3′.
- Point $P1$ is located the distance D from the lateral edge 3-3′ and is transferred from the top view to the development. Points $P2$ and $P4$ are transferred in a similar manner.
- Connect these points with straight lines and indicate on the development that the area enclosed is to be a "cut-out."

The development described is typical of ductwork made out of sheet metal. Lateral edge 1-1′ was selected as the seam to give extra strength to the segment of ductwork to be fabricated from this pattern. Usually, extra material is allowed at both ends of the pattern in case it is desired to crimp the seam joint as indicated in the

77

inset. Lateral edges 2-2', 3-3', and 4-4' are called *bend lines* and they are usually scored with a pointed instrument called a *scribe* to facilitate bending during the fabrication of the part.

The pattern is developed with the inside up and it is customary to indicate this fact on the drawing. It is easier to bend the sides of a sheet metal part up, at the layout table, than it is to bend them down.

5.13 DEVELOPMENT OF AN OBLIQUE PRISM

The triangular prism in Figure 5-4 is an oblique prism. The development of an oblique prism is no different than the development of a right prism. The oblique triangular prism has been reoriented perpendicular to the horizontal plane in Figure 5-15. It is now a right triangular prism and is developed as in Section 5.12.

Figure 5-14 Development of a Right Square Prism

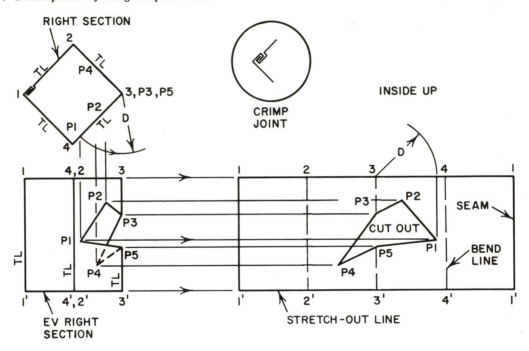

Figure 5-15 Development of a Triangular Prism

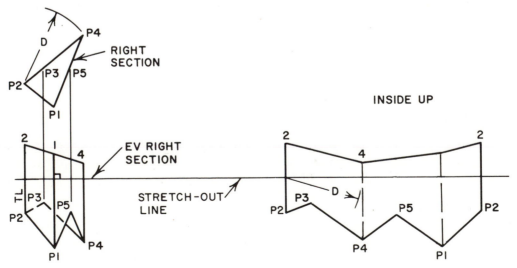

78

Figure 5-16 Development of a Right Circular Cylinder

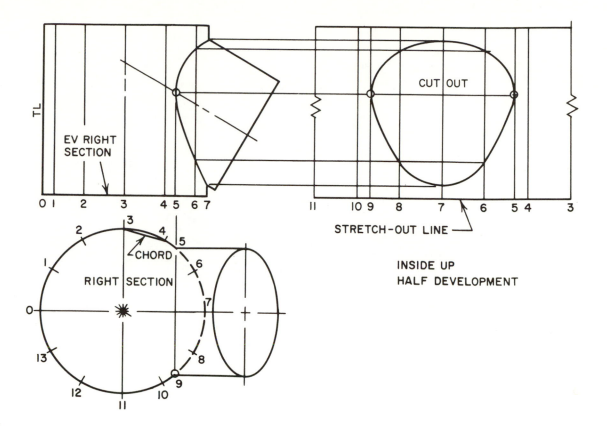

5.14 DEVELOPMENT OF A RIGHT CIRCULAR CYLINDER

A right circular cylinder intersected by an oblique circular cylinder is shown in Figure 5-16. The base of the right circular cylinder has been divided into twelve equal parts. The stretch-out line is projected from the edge view of the right section. Chord lengths taken from the right section are transferred to the stretch-out line. This procedure is the same as that used for development of a right prism.

The chord is slightly less than the arc it subtends and will introduce some error into the development. Dividing the circle into twelve parts provides sufficient accuracy.

The two points 5 and 9 on the base locate limiting elements on the line of intersection between the right cylinder and the oblique cylinder. Points 5 and 9 can be located on the stretch-out line by taking the chordal distances 6-5 and 8-9, respectively, from the right section. The cylinder elements are true length in the view which shows the edge view of the right section and are projected, along with the points on the curve of intersection, to the development.

5.15 DEVELOPMENT OF AN OBLIQUE CYLINDER

The development of an oblique cylinder requires first and second auxiliary views to determine the true length of the elements and the true shape of a right section. Figure 5-17 illustrates this development.

The straight line elements used in developments of prisms and cylinders are true length and parallel in the view that shows the right section on edge. Such developments are generally classified as parallel-line developments.

5.16 RADIAL LINE DEVELOPMENTS

Pyramid and cone developments, which will be discussed next, are called *radial line developments* because they are developed with straight line elements that converge at a common point. A single view of a pyramid or cone does not show the true length of all the straight line elements needed in the development. Rotation is normally used to find the necessary true lengths, and the developments may be started at any convenient position.

79

Figure 5-17 Development of an Oblique Cylinder

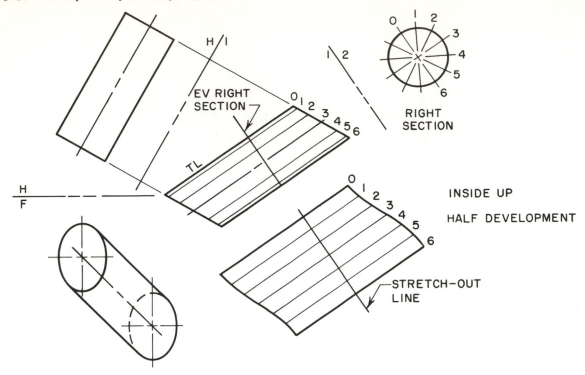

5.17 DEVELOPMENT OF A RIGHT PYRAMID

A truncated right pyramid with a square base is shown in Figure 5-18. It is generally desirable to start the development on the shortest seam for the sake of economy, particularly if welding or soldering is involved in fabricating the part. The short seam 0-5 is on a frontal line and is a part of the straight line element V-0, which is true length in the front view.

- Draw V-0 in a convenient position to start the development and then locate point 5 on V-0. The true length of 0-1 is shown in the top view.
- Use point 0 on the development as a center and swing an arc of length 0-1. Find the true length of element V-1 by rotation.
- From the vertex V swing an arc V-1 to cut the arc 0-1.
- Extend the arc of radius V-1 as shown because V-2, V-3, and V-4 are the same length as V-1.
- Using the true length of edge 1-2 from the top or front view as a radius and point 1 as a center, swing an arc until it cuts the previous arc.
- Repeat this operation to locate points 3 and 4 on the development.
- Find point 0 to return to the original starting point for the development.
- The true-length distances for locating points 6, 7, 8,

and 9 are obtained from the front view after being rotated.

Note that lines on a face of the pyramid that are parallel or perpendicular also have the same relationship on the development. This is a good way to check for accuracy.

5.18 DEVELOPMENT OF AN OBLIQUE PYRAMID

None of the radial straight line edges of the truncated rectangular oblique pyramid in Figure 5-19 is true length in the given views. The true lengths of the radial lines are obtained by rotation. See the TL diagram.

- The true length of V-0 is drawn in a convenient position to start the development. The distance for locating point 4 on V-0 is also taken from the *TL* diagram.
- Swing an arc from point 0 using as a radius the true length of edge 0-1 taken from the top view.
- Swing another arc from V using the true length of V-1 taken from the *TL* diagram. These arcs cross to locate point 1 on the development.
- Continue in a similar fashion until the other points on the base have been located.
- Parallelism has been utilized to complete the development.

- Select a convenient location for the vertex V in the development and draw an arc with radius V-0.
- Starting at point 0 on the development, strike twelve

arcs with a radius 0–1 to locate all the points on the circular base.

Figure 5–18 **Development of a Right Pyramid**

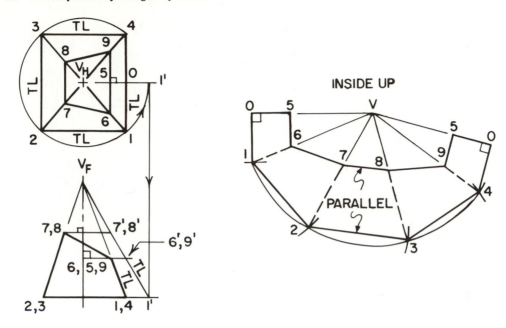

Figure 5–19 **Development of an Oblique Pyramid**

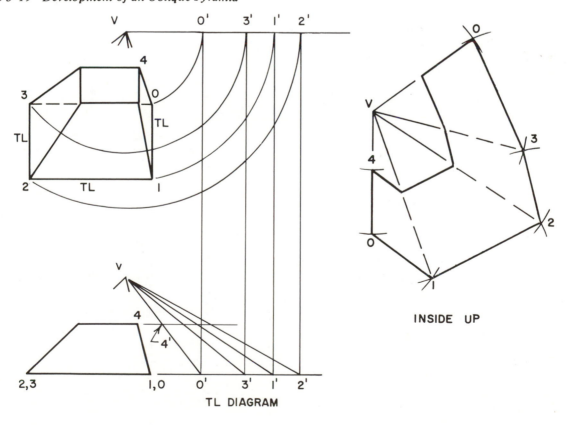

Figure 5-20 Development of a Right Circular Cone

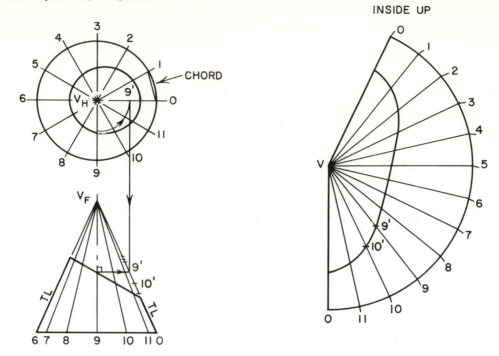

Figure 5-21 Development of an Oblique Cone

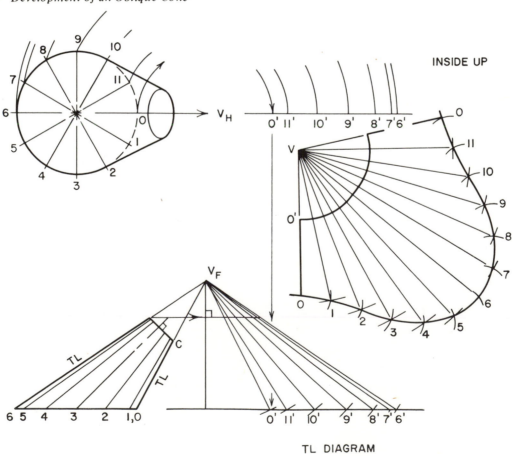

TL DIAGRAM

82

5.19 DEVELOPMENT OF A RIGHT CIRCULAR CONE

The development of a truncated right circular cone is shown in Figure 5-20. All elements of a right circular cone are the same length. The element V-0 is typical, and its true length is obtained from the front view. Also, the chordal distances are equal, and the true length of the typical chord 0-1 is obtained from the top view.

- Select a convenient location for the vertex V in the development and draw an arc with radius V-0.
- Starting at point 0 on the development, strike twelve arcs with a radius 0-1 to locate all the points on the circular base.
- Find the true-length distances from the vertex to each of the points on the inclined surface. This is done by rotation as shown in the given views for a typical point 9.
- The true-length distance from the vertex to the point is found in the front view and transferred to the development.
- It is not necessary to draw the circular arcs in the top view. Perpendiculars to the axis in the front view are sufficient to find the true lengths along the element V_F-0 as shown in the illustration (V_F-10').

5.20 DEVELOPMENT OF AN OBLIQUE CONE

The development of the truncated oblique cone shown in Figure 5-21 is similar to the preceding development, except that the true length of each element is different. A *TL* diagram is drawn to obtain the true lengths of all the elements.

Although any right section of the cone is elliptical, the base is a circle, which is true shape in the top view. The chordal distances of the base are equal and show true length in the top view. The true-length distances from the vertex V to the elliptical opening are obtained by rotation as shown in the *TL* diagram.

5.21 DEVELOPMENT OF TRANSITION PIECES

Transition pieces are used whenever it is necessary to change from an opening of one geometric shape to another. The method used in developing transition pieces is called *triangulation.*

The transition piece shown in Figure 5-22 has a square opening in the top and a rectangular opening in the bottom.

- The development is started on the short seam 0-9. The panel 0-1-5-9 is true shape in the front view and is transferred in reverse so that the development will be inside up.
- Edges 1-2 and 5-6 are true length in the top view, and it should be noted that they are perpendicular to bend line 1-5. The lines 1-2 and 5-6 are added to the development.
- The panel 1-2-6-5 is completed by connecting points 2 and 6 to obtain another bend line.
- Triangulation, instead of a perpendicularity relationship, must be used to determine the development of panel 2-3-7-6.

Figure 5-22 Development of a Transition Piece

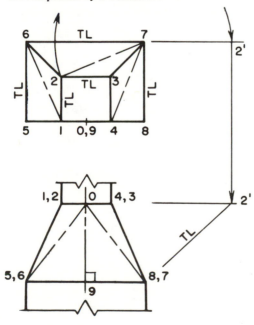

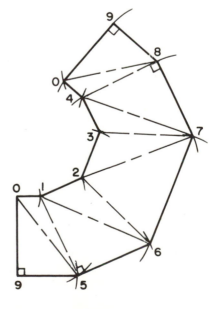

INSIDE UP

83

Figure 5-23 Development of a Transition Piece

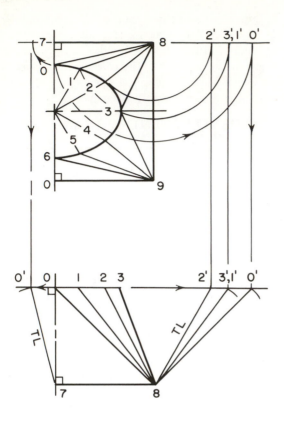

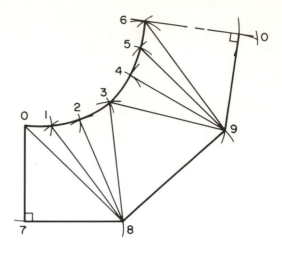

INSIDE UP

HALF DEVELOPMENT

- The true length of the diagonal 2-7 is found by rotation. Using point 2 on the development as a center, swing an arc whose length is equal to 2-7.
- Using the *TL* of edge 6-7 from the top view as a radius and point 6 on the development as a center, swing an arc intersecting the previous arc to locate point 7 by triangulation.
- Edge 2-3 is drawn parallel to edge 6-7. The bend line 3-7 completes the panel 2-3-7-6.
- Since the transition piece is symmetrical, the remaining two panels are constructed as the reverse of panels 1-2-6-5 and 0-1-5-9.

The transition piece shown in Figure 5-23 has an elliptical opening in the top and a rectangular opening in the bottom. To develop this transition piece, it is divided into cone segments and triangles as shown in the given views. The chordal distances 0-1, 1-2, and 2-3, seen in true length in the top view, are not equal, and care must be exercised when these chords are used in the development. The development is constructed by triangulation, and the accuracy of the development can be assured if perpendicularity is attained at the points so indicated on the development. Half-views were used because the transition piece is symmetrical. The half-development can be turned over to make a full pattern.

Technical Practices

6

Technical practices include the drawing of section views, basic dimensioning, tolerancing, and representation of threaded fasteners. Section views are often more descriptive than exterior views. Dimensioning and tolerancing add the size description to the basic shape description of an object. Threaded fasteners are used in most aspects of engineering design. This section shows the more common threaded symbols.

6.1 SECTION VIEWS

Drawings of objects may be simplified by the use of *section views*, in which part of the object is removed to expose interior details. Figure 6-1 illustrates a full section view. The right half of the object has been removed where an imaginary vertical cutting plane A-A cuts through the center of the object, as indicated by the cutting plane line A-A in the front view. The material cut by the cutting plane is shown by the section-lined areas in the sectioned view. The arrows on the cutting plane line show the direction of sight.

The *section lines* are drawn as thin evenly spaced parallel lines spaced 2 to 3 mm apart and sloped up to the right at 45°. In assembly sections the angle of the section lines is changed for different parts for ease of identification. See Figure 7-4. Figure 6-2 illustrates four standard section lining or "crosshatching" symbols. Very thin sections are shaded solid black. See Figure 7-6.

Figure 6-2   Typical Section-Line Symbols

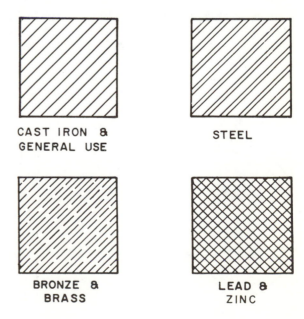

CAST IRON & GENERAL USE

STEEL

BRONZE & BRASS

LEAD & ZINC

Figure 6-1   Section Views

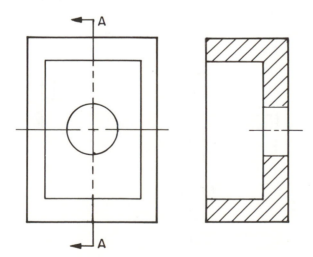

6.2 TYPES OF SECTIONS

A *full section* is a view obtained by passing the cutting plane entirely through the object either in a straight line or an offset line to include important features. All visible features are shown, whether in the section plane or beyond it. Invisible features are omitted unless essential for clarity. In Figure 6-1 the cutting plane passes straight through the object.

In Figure 6-3 the plane is offset to pass through the round end slot. Note that changes in direction of the cutting plane are not shown in the section-lined areas.

Figure 6-3 Offset Section

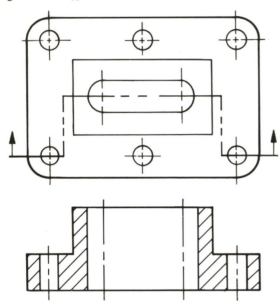

A *half-section* is a view obtained by removing *one-fourth of the object*. It is useful for symmetrical objects because it combines the advantages of both section and exterior views. Hidden lines are usually omitted in the two halves that join in a common center line. See Figure 6-4.

Figure 6-4 Half-Section

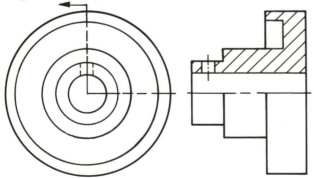

A *broken-out section* is a partial section used to show an interior detail by breaking away less than half of the object. An irregular break line is used to show the extent of the break. Hidden lines are usually necessary in this type of section as illustrated in Figure 6-5.

Figure 6-5 Broken-Out Section

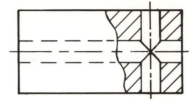

A *revolved section* is a section placed on a conventional view to show the true shape of an element such as a spoke, arm, or shaft. In theory the imaginary cutting plane is placed perpendicular to the longitudinal axis of the element and then revolved 90°. Figure 6-6 illustrates revolved sections.

Figure 6-6 Revolved Sections

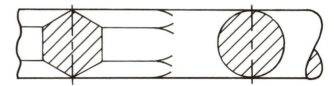

A *removed section* is a section view moved from its projected position to any convenient location on the drawing sheet. It may be a partial or a complete view and must be identified as section A-A, B-B, etc., corresponding to the letters at the arrows on the cutting-plane line. Figure 6-7 illustrates a partial top view and a removed section of the bracket showing the rounded base.

Figure 6-7 Removed Section

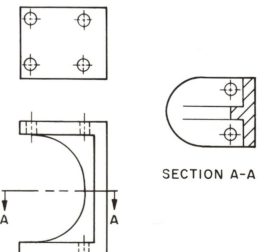

SECTION A-A

Figure 6-8 Sectioning

Figure 6-9 Sectioning

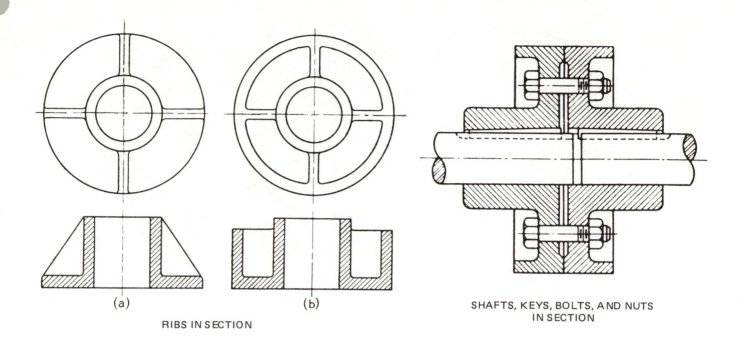

(a)   (b)

RIBS IN SECTION

SHAFTS, KEYS, BOLTS, AND NUTS
IN SECTION

6.3 CONVENTIONS

Some conventional practices are violations of true projection which make a drawing easier to visualize. Elements such as thin ribs, spokes, keys, pins, shafts, rivets, bolts and other solid parts that lie in the section plane are not sectioned. This is done to avoid a false impression of thickness or solidity. See Figures 6-8 and 6-9.

Another convention is the rotation, or alignment into the plane of the paper, of elements such as holes, ribs, spokes, etc., to simplify a view that otherwise would be time consuming to draw and difficult to read. In Figure 6-10, a spoke, a hole, and rib have been rotated to make the section views symmetrical. Figure 6-11 illustrates the simplifying aspects of this convention.

Figure 6-10 Rotation of Elements

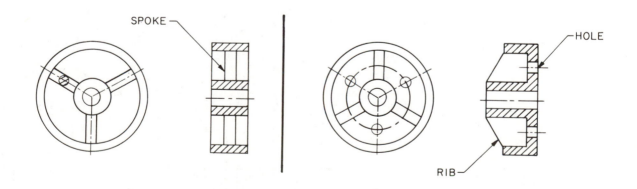

Figure 6-11 Spokes in Section

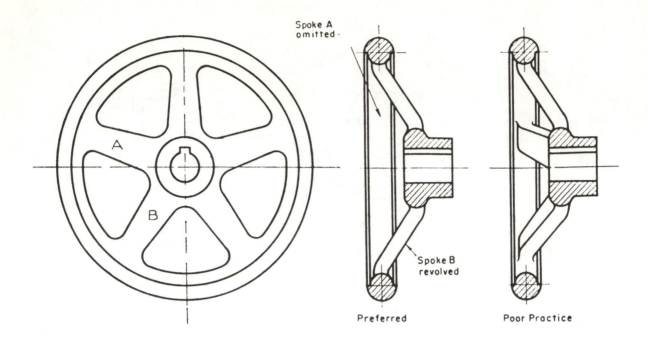

Spoke A
omitted

Spoke B
revolved

Preferred   Poor Practice

Sections through an intersection where the exact lines of intersection are very small or of no consequence may be simplified by using the convention shown in Figures 6-12(a) and (c). Larger figures of intersection may be exactly projected as shown in Figure 6-12(b), or approximated by circular arcs as in Figure 6-12(d).

Figure 6-12 Intersections in Section

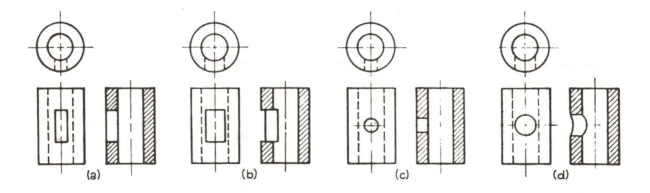

(a)   (b)   (c)   (d)

6.4 BASIC DIMENSIONING

The complete description of an object includes the shape, size, and location of the components relative to each other. The shape description is shown by the proper use of principal and auxiliary views. The size and location description are shown by the proper use of dimensions.

Dimensioning includes extension lines, dimension lines, arrowheads, and notes, as well as the actual dimension numerals. Figure 6-13 is an example of a correctly dimensioned object and illustrates the proper spacing and placement of the several dimensioning components. Extension lines, dimension lines, and leaders are thin black lines of the same line weight as center lines and section lines. A space of about 2 mm is left between an object line and an extension line unless a center line is used as an extension line, in which case the center line may cross the object line.

Dimensions should be placed outside views and between views whenever possible. The first dimension should be placed approximately 10 mm from the object line; succeeding dimensions are about 8 mm apart. Arrowheads are about 3 mm long and should be thin, neat, and uniform.

Notes are always lettered horizontally and guidelines for 3 mm high letters are used. Dimensions are the same height and may be lettered without guidelines if the draftsman is particularly adept at lettering.

Dimensions may be placed so that all are read from the bottom of the page, in which case they are called *unidirectional dimensions*. Dimension numerals placed in line with dimension lines are referred to as *aligned dimensions*. Figure 6-13 has aligned dimensions.

Figure 6-13 Placement of Dimensions

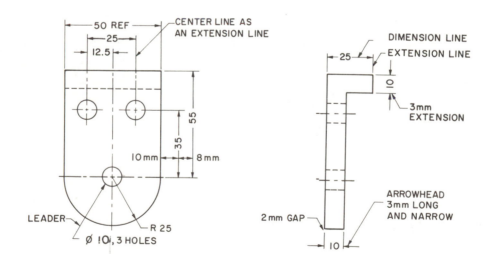

6.5 DIMENSIONING BASIC SHAPES

The basic geometric shapes most used in engineering design are the cylinder, cone, prism, pyramid, and sphere. In the illustrative figures the letters S (size) and L (location) are used in place of the actual dimensions. A right circular cylinder is dimensioned by giving the height or length and the diameter of the cylinder. Both dimensions are applied to the rectangular view as shown in Figure 6-14(a). Less than half of a right circular cylinder would be dimensioned as shown in Figure 6-14(b).

A right circular cone is dimensioned by giving the height and base diameter. See Figure 6-14(c). The frustum of a cone is dimensioned as shown in Figure 6-14(d).

A right prism with a rectangular base is dimensioned as shown in Figure 6-14(e). Since a prism may have any polygon for a base, it may be necessary to use a combination of linear and angular dimensions to indicate the size of the base, as illustrated in Figure 6-14(f).

A frustum of a right rectangular pyramid is dimensioned as shown in Figure 6-14(g) and a sphere is dimensioned by giving the diameter as shown in Figure 6-14(h).

Figure 6-14   Dimensioning Basic Shapes

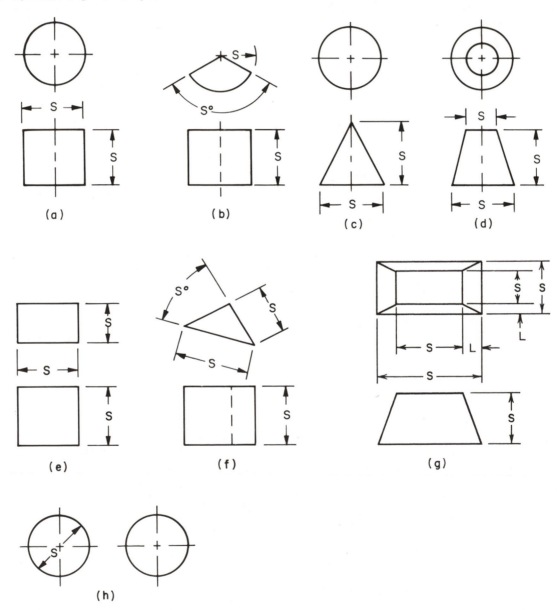

90

Holes are sometimes referred to as *negative shapes*, and similar solids are referred to as *positive shapes*. Cylindrical holes are dimensioned with a note in the view showing the hole as a circle. This note gives the diameter of the hole. The note also specifies the number of holes of the same size, if this fact is fairly obvious on the drawing. See Figure 6-16. A rounded corner is indicated by the radius.

A *leader* drawn at a convenient angle is directed toward the center of a hole or arc with the arrowhead touching the arc. Leaders are also used to connect notes with corresponding features. See Figure 6-13.

Overall dimensions extending from one end of the part to the other are usually given except for parts which have rounded ends, as shown in Figure 6-13. In these cases, the radius must be added to the location dimensions to get the overall height. The radius is also the primary dimension for the width of the part. The reference dimension, 50 REF, is a secondary dimension and is added if the designer believes that a secondary dimension would be helpful to the person reading the drawing.

6.6 LOCATION DIMENSIONS

Location dimensions position the basic geometric shapes with respect to each other. Whenever possible, these dimensions should be given from center lines and machined surfaces in the views showing the characteristic shapes. Machining operations not specified by notes are indicated by *finish marks*. The finish mark is placed on <u>all</u> edge views of a machined surface. The finish mark is a "√" drawn as shown in Figure 6-15.

Sometimes a dimension may give the size as well as the location, as indicated by the double notation S-L in Figure 6-16. The positive cylinder on the left end of the base prism is located from the left rear corner in the top view by two location dimensions. The holes, one near the center of the base and the other on the right side, are located by dimensions between center lines. The center of the rounded corner is self-locating, in this case by the radius.

Figure 6-16 Location Dimensions

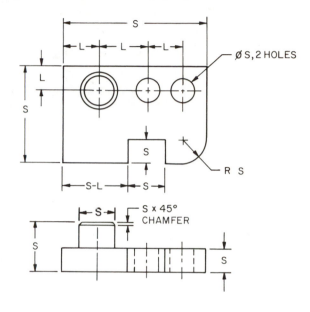

Figure 6-15 Fillets and Rounds

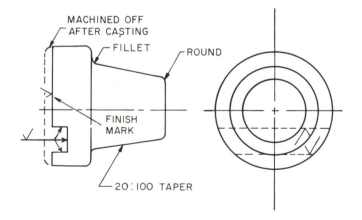

6.7 CROWDED DIMENSIONS

The shape of an object may be such that a limited space is available for dimensioning. Figure 6-17 illustrates how crowded dimensions may be treated. Other situations not illustrated will likely occur, but the experienced designer is usually able to work out a satisfactory solution.

Figure 6-17 Crowded Dimensions

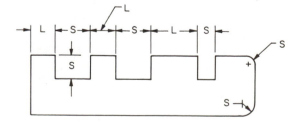

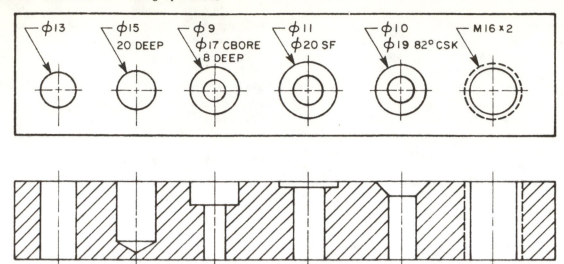

Figure 6-18 Small-Tool Machining Operations

6.8 FILLETS AND ROUNDS

Parts produced by casting material in a mold generally have internally rounded corners called *fillets* or externally rounded corners called *rounds*. See the notes on Figure 6-15. The practice of rounding corners is followed to prevent shrinkage cracks that might appear in corners or on edges after the part is cooled. A slight taper may also be added to make it easier to remove a part from its mold.

6.9 SMALL-TOOL MACHINING OPERATIONS

Most of the small-tool machining operations performed in the machine shop are illustrated in Figure 6-18. In each case a hole is produced using a high-speed twist drill that is cone-shaped on the end for penetration of the material. A drill has fluted spiral cutting edges producing the proper diameter. A counterbore is a tool used for enlarging part of the drilled hole for recessing a fillister head bolt or screw. A spotface is similar in appearance to a counterbore and usually cuts just deep enough to produce a machined bearing surface for the head of a bolt or a nut. A countersink tool in the shape of a cone is used to recess material to fit the head of a flathead or ovalhead screw. A tapped hole is drilled first and then threaded with a fluted tool called a *tap*, which cuts the threads. Dimensions for these machined holes are generally given in note form.

6.10 RULES OF DIMENSIONING

There are many rules governing dimensioning which are accepted as standard practice. They should be followed

whenever possible. The most important of these basic rules are

1. Dimension the view that best describes the shape of the feature.
2. Place dimensions outside the boundary of the views where possible and as near the feature as spacing permits.
3. Locate holes in the circular views by dimensioning to center lines.
4. Dimension the size of a hole by its diameter on the circular view using a leader.
5. Dimension the diameter of a positive cylinder or cone in the noncircular view.
6. Dimension the radius of an arc in its true shape view.
7. Avoid crowding dimensions.
8. Avoid cumulative errors by giving the overall dimension and omitting one dimension in a series of dimensions.
9. Avoid crossing dimension lines and extension lines.
10. Avoid dimensioning to hidden lines.
11. Never use a center line, an object line, or an extension line as a dimension line.

6.11 LIMIT DIMENSIONS

Manufactured parts must, in many cases, be interchangeable, irrespective of where or how they are produced. Additionally, the size of parts which fit and function together must be controlled so that they will perform as the designer intends. A method of specifying dimensions of such "mating" parts is known as *limit dimensioning*.

Limit dimensioning involves the selection of appropriate "tolerance" in the actual production of a part and the selection of appropriate "allowance" in the actual fitting together of mating parts.

Tolerance is the permissible variation in the size of a part, that is, the difference between its maximum and minimum sizes.

Allowance is the specified size difference between mating parts which provides for the tightest fit. A *positive allowance* provides a minimum clearance between mating parts. *Negative allowance* specifies the maximum interference in the fit of mating parts.

Some additional definitions pertinent to limit dimensioning are as follows.

1. *Nominal size:* approximately actual size—used for general identification.
2. *Basic size:* the theoretical size from which limit dimensions are obtained by application of appropriate tolerance and allowance.
3. *Clearance fit:* mating such that there is *always* clearance between parts (for example, a shaft turning freely in a bearing).
4. *Transition fit:* limit dimensions such that either a clearance or an interference may result when parts are mated.
5. *Interference fit:* mating such that there is always interference between parts (for example, a pin forced into a hole).

6. *Basic hole system:* limit dimensions of mating parts are computed by applying tolerance and allowance to the *minimum size* of the hole. This is a system of limit dimensioning in which the minimum size of the hole is the basic size to which tolerance and allowance are applied for both the hole and the shaft.
7. *Basic shaft system:* limit dimensions of mating parts are computed by applying tolerance and allowance to the *maximum size* of the shaft. This is a system of limit dimensioning in which the maximum size of the shaft is the basic size to which tolerance and allowance are applied for both parts.

6.12 LIMIT DIMENSIONS—MATING PARTS

Either the basic hole system or the basic shaft system may be used in developing limit dimensions for mating parts. From a manufacturing viewpoint, the basic hole system is preferred. However, when a common shaft is mated with several holes, the basic shaft system should be used.

Metric standards for tolerance and allowance have been developed for the American National Standards Institute (ANSI) document B4.2, "Preferred Metric Limits and Fits." The document defines ten *preferred* fits for both the basic hole and basic shaft systems, as prescribed by the ISO (International Organization for Standardization). A description of these preferred fits is contained in Figure 6-19. Other fits may be used by designers, but they must then calculate the dimensions from various tables.

Figure 6-19 Description of Preferred Metric Fits

	ISO SYMBOL		
	Hole Base	Shaft Base	DESCRIPTION
Clearance Fits	H11/c11	C11/h11	Loose running fit for wide commercial tolerances or allowance on external members.
	H9/d9	D9/h9	Free running fit not for use where accuracy is essential, but good for large temperature variations, high running speeds, or heavy journal pressures.
	H8/f7	F8/h7	Close running fit for running on accurate machines and for accurate location at moderate speeds and journals pressures.
	H7/g6	G7/h6	Sliding fit not intended to run freely, but to move and turn freely and locate accurately.
	H7/h6	H7/h6	Locational clearance fit provides snug fit for locating stationary parts, but can be freely assembled and disassembled.
Transition Fits	H7/k6	K7/h6	Locational transition fit for accurate location; a compromise between clearance and interference.
	H7/n6	N7/h6	Locational transition fit for more accurate location where greater interference is permissible.
Interference Fits	H7/p6	P7/h6	Locational interference fit for parts requiring rigidity and alignment with prime accuracy of location but without special bore pressure requirements.
	H7/s6	S7/h6	Medium drive for ordinary steel parts or shrink fits on light sections; the tightest fit usable with cast iron.
	H7/u6	U7/h6	Force fit suitable for parts which can be highly stressed or for shrink fits where the heavy pressing forces required are impractical.

More Clearance — More Interference

93

Figure 6-20 Preferred Metric Sizes

NOMINAL SIZE (mm)		NOMINAL SIZE (mm)		NOMINAL SIZE (mm)	
FIRST	SECOND	FIRST	SECOND	FIRST	SECOND
1		10		100	
	1.1		11		110
1.2		12		120	
	1.4		14		140
1.6		16		160	
	1.8		18			180
2		20		200	
	2.2		22		220
2.5		25		250	
	2.8		28		280
3		30		300	
	3.5		35			350
4		40		400	
	4.5		45		450
5		50		500	
	5.5		55		550
6		60		600	
	7		70		700
8		80		800	
	9		90		900
				1000	

Preferred metric sizes for internal or external parts are given in Figure 6-20. Preferred tolerance for holes and shafts are given for a range of sizes in Tables 6-1 and 6-2 excerpted from ANSI B4.2.

Figure 6-21 illustrates the selection of limit dimensions for clearance, transition, and interference fits using the basic hole system. The clearance fit (selected from Figure 6-19) is designated at 25 H8/f7 and will provide a close running fit. In this designation the "25" is the basic size of the hole in millimeters and also the minimum hole diameter. The capital letter "H" coupled with the number "8" refers to the tolerance zone applied to the internal dimension (the hole) using the basic hole system. The lowercase letter "f" coupled with the number "7" defines the tolerance zone applied to the external dimension (the shaft). The difference between the minimum hole diameter and the maximum shaft diameter is the allowance of 0.020 mm. Actual limit dimensions for this clearance fit are given in Table 6-3.

Had the basic shaft system been used, the fit designation for the same clearance would be 25 F8/h7 (see Figure 6-19). In this case, the "25" is the basic size of the shaft and the maximum shaft diameter. The allowance is again 0.020 mm.

Tolerance applied to a part may be indicated as shown in any one of the following examples, which pertain to a 25 mm basic diameter hole.

$$25\text{H8} \qquad 25\text{H8} \left(\begin{matrix} 25.033 \\ 25.000 \end{matrix} \right)$$

$$25 \begin{matrix} +0.033 \\ -0.000 \end{matrix} \qquad \frac{25.033}{25.000}$$

Similarly, preferred transition and interference fits are indicated using the basic hole system. Actual limit dimensions for these transition and interference fits are taken from Table 6-4.

Figure 6-21   Limit Dimensions

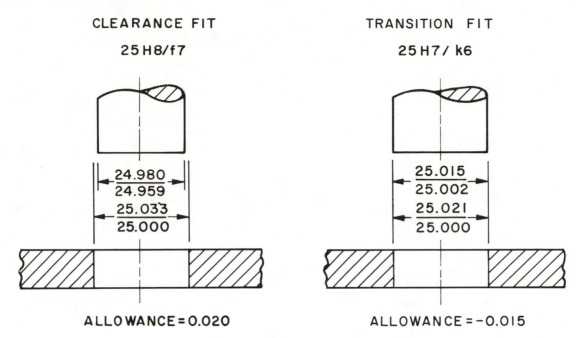

CLEARANCE FIT
25 H8/f7

24.980
24.959

25.033
25.000

ALLOWANCE = 0.020

TRANSITION FIT
25 H7/ k6

25.015
25.002

25.021
25.000

ALLOWANCE = -0.015

INTERFERENCE FIT
25 H7 / s6

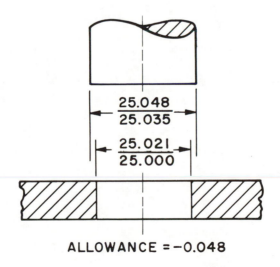

25.048
25.035

25.021
25.000

ALLOWANCE = -0.048

NOMINAL DIAMETER = 25 mm   DIMENSIONS IN MILLIMETERS

Table 6-1 First Choice Tolerance Zones—Holes

Dimensions in Millimeters

SIZE		C11	D9	F8	G7	H11	H9	H8	H7	K7	N7	P7	S7	U7
FROM	1	+0.120	+0.045	+0.020	+0.012	+0.060	+0.025	+0.014	+0.010	0.000	−0.004	−0.006	−0.014	−0.018
TO	3	+0.060	+0.020	+0.006	+0.002	0.000	0.000	0.000	0.000	−0.010	−0.014	−0.016	−0.024	−0.028
OVER	3	+0.145	+0.060	+0.028	+0.016	+0.075	+0.030	+0.018	+0.012	+0.003	−0.004	−0.008	−0.015	−0.019
TO	6	+0.070	+0.030	+0.010	+0.004	+0.000	0.000	0.000	0.000	−0.009	−0.016	−0.020	−0.027	−0.031
OVER	6	+0.170	+0.076	+0.035	+0.020	+0.090	+0.036	+0.022	+0.015	+0.005	−0.004	−0.009	−0.017	−0.022
TO	10	+0.080	+0.040	+0.013	+0.005	0.000	0.000	0.000	0.000	−0.010	−0.019	−0.024	−0.032	−0.037
OVER	10	+0.205	+0.093	+0.043	+0.024	+0.110	+0.043	+0.027	+0.018	+0.006	−0.005	−0.011	−0.021	−0.026
TO	14	+0.095	+0.050	+0.016	+0.006	0.000	0.000	0.000	0.000	−0.012	−0.023	−0.029	−0.039	−0.044
OVER	14	+0.205	+0.093	+0.043	+0.024	+0.110	+0.043	+0.027	+0.018	+0.006	−0.005	−0.011	−0.021	−0.026
TO	18	+0.095	+0.050	+0.016	+0.006	0.000	0.000	0.000	0.000	−0.012	−0.023	−0.029	−0.039	−0.044
OVER	18	+0.240	+0.117	+0.053	+0.028	+0.130	+0.052	+0.033	+0.021	+0.006	−0.007	−0.014	−0.027	−0.033
TO	24	+0.110	+0.065	+0.020	+0.007	0.000	0.000	0.000	0.000	−0.015	−0.028	−0.035	−0.048	−0.054
OVER	24	+0.240	+0.117	+0.053	+0.028	+0.130	+0.052	+0.033	+0.021	+0.006	−0.007	−0.014	−0.027	−0.040
TO	30	+0.110	+0.065	+0.020	+0.007	0.000	0.000	0.000	0.000	−0.015	−0.028	−0.035	−0.048	−0.061
OVER	30	+0.280	+0.142	+0.064	+0.034	+0.160	+0.062	+0.039	+0.025	+0.007	−0.008	−0.017	−0.034	−0.051
TO	40	+0.120	+0.080	+0.025	+0.009	0.000	0.000	0.000	0.000	−0.018	−0.033	−0.042	−0.059	−0.076
OVER	40	+0.290	+0.142	+0.064	+0.034	+0.160	+0.062	+0.039	+0.025	+0.007	−0.008	−0.017	−0.034	−0.061
TO	50	+0.130	+0.080	+0.025	+0.009	0.000	0.000	0.000	0.000	−0.018	−0.033	−0.042	−0.059	−0.086
OVER	50	+0.330	+0.174	+0.076	+0.040	+0.190	+0.074	+0.046	+0.030	+0.009	−0.009	−0.021	−0.042	−0.076
TO	65	+0.140	+0.100	+0.030	+0.010	0.000	0.000	0.000	0.000	−0.021	−0.039	−0.051	−0.072	−0.106
OVER	65	+0.340	+0.174	+0.076	+0.040	+0.190	+0.074	+0.046	+0.030	+0.009	−0.009	−0.021	−0.048	−0.091
TO	80	+0.150	+0.100	+0.030	+0.010	0.000	0.000	0.000	0.000	−0.021	−0.039	−0.051	−0.078	−0.121
OVER	80	+0.390	+0.207	+0.090	+0.047	+0.220	+0.087	+0.054	+0.035	+0.010	−0.010	−0.024	−0.058	−0.111
TO	100	+0.170	+0.120	+0.036	+0.012	0.000	0.000	0.000	0.000	−0.025	−0.045	−0.059	−0.093	−0.146
OVER	100	+0.400	+0.207	+0.090	+0.047	+0.220	+0.087	+0.054	+0.035	+0.010	−0.010	−0.024	−0.066	−0.131
TO	120	+0.180	+0.120	+0.036	+0.012	0.000	0.000	0.000	0.000	−0.025	−0.045	−0.059	−0.101	−0.166
OVER	120	+0.450	+0.245	+0.106	+0.054	+0.250	+0.100	+0.063	+0.040	+0.012	−0.012	−0.028	−0.077	−0.155
TO	140	+0.200	+0.145	+0.043	+0.014	0.000	0.000	0.000	0.000	−0.028	−0.052	−0.068	−0.117	−0.195
OVER	140	+0.460	+0.245	+0.106	+0.054	+0.250	+0.100	+0.063	+0.040	+0.012	−0.012	−0.028	−0.085	−0.175
TO	160	+0.210	+0.145	+0.043	+0.014	0.000	0.000	0.000	0.000	−0.028	−0.052	−0.068	−0.125	−0.215
OVER	160	+0.480	+0.245	+0.106	+0.054	+0.250	+0.100	+0.063	+0.040	+0.012	−0.012	−0.028	−0.093	−0.195
TO	180	+0.230	+0.145	+0.043	+0.014	0.000	0.000	0.000	0.000	−0.028	−0.052	−0.068	−0.133	−0.235
OVER	180	+0.530	+0.285	+0.122	+0.061	+0.290	+0.115	+0.072	+0.046	+0.013	−0.014	−0.033	−0.105	−0.219
TO	200	+0.240	+0.170	+0.050	+0.015	+0.000	0.000	0.000	0.000	−0.033	−0.060	−0.079	−0.151	−0.265
OVER	200	+0.550	+0.285	+0.122	+0.061	+0.290	+0.115	+0.072	+0.046	+0.013	−0.014	−0.033	−0.113	−0.241
TO	225	+0.260	+0.170	+0.050	+0.015	0.000	0.000	0.000	0.000	−0.033	−0.060	−0.079	−0.159	−0.287
OVER	225	+0.570	+0.285	+0.122	+0.061	+0.290	+0.115	+0.072	+0.046	+0.013	−0.014	−0.033	−0.123	−0.267
TO	250	+0.280	+0.170	+0.050	+0.015	0.000	0.000	0.000	0.000	−0.033	−0.060	−0.079	−0.169	−0.313
OVER	250	+0.620	+0.320	+0.137	+0.069	+0.320	+0.130	+0.081	+0.052	+0.016	−0.014	−0.036	−0.138	−0.295
TO	280	+0.300	+0.190	+0.056	+0.017	0.000	0.000	0.000	0.000	−0.036	−0.066	−0.088	−0.190	−0.347
OVER	280	+0.650	+0.320	+0.137	+0.069	+0.320	+0.130	+0.081	+0.052	+0.016	−0.014	−0.036	−0.150	−0.330
TO	315	+0.330	+0.190	+0.056	+0.017	0.000	0.000	0.000	0.000	−0.036	−0.066	−0.088	−0.202	−0.382
OVER	315	+0.720	+0.350	+0.151	+0.075	+0.360	+0.140	+0.089	+0.057	+0.017	−0.016	−0.041	−0.169	−0.369
TO	355	+0.360	+0.210	+0.062	+0.018	0.000	0.000	0.000	0.000	−0.040	−0.073	−0.098	−0.226	−0.426
OVER	355	+0.760	+0.350	+0.151	+0.075	+0.360	+0.140	+0.089	+0.057	+0.017	−0.016	−0.041	−0.187	−0.414
TO	400	+0.400	+0.210	+0.062	+0.018	0.000	0.000	0.000	0.000	−0.040	−0.073	−0.098	−0.244	−0.471
OVER	400	+0.840	+0.385	+0.165	+0.083	+0.400	+0.155	+0.097	+0.063	+0.018	−0.017	−0.045	−0.209	−0.467
TO	450	+0.440	+0.230	+0.068	+0.020	0.000	0.000	0.000	0.000	−0.045	−0.080	−0.108	−0.272	−0.530
OVER	450	+0.880	+0.385	+0.165	+0.083	+0.400	+0.155	+0.097	+0.063	+0.018	−0.017	−0.045	−0.229	−0.517
TO	500	+0.480	+0.230	+0.068	+0.020	0.000	0.000	0.000	0.000	−0.045	−0.080	−0.108	−0.292	−0.580

Tables 6-1 through 6-6 excerpted from ANSI B4.2 "Preferred Metric Limits and Fits."

Table 6–2 First Choice Tolerance Zones—Shafts

Dimensions in Millimeters

SIZE		c11	d9	f7	g6	h11	h9	h7	h6	k6	n6	p6	s6	u6
FROM	1	−0.060	−0.020	−0.006	−0.002	0.000	0.000	0.000	0.000	+0.006	+0.010	+0.012	+0.020	+0.024
TO	3	−0.120	−0.045	−0.016	−0.008	−0.060	−0.025	−0.010	−0.006	0.000	+0.004	+0.006	+0.014	+0.018
OVER	3	−0.070	−0.030	−0.010	−0.004	0.000	0.000	0.000	0.000	+0.009	+0.016	+0.020	+0.027	+0.031
TO	6	−0.145	−0.060	−0.022	−0.012	−0.075	−0.030	−0.012	−0.008	+0.001	+0.008	+0.012	+0.019	+0.023
OVER	6	−0.080	−0.040	−0.013	−0.005	0.000	0.000	0.000	0.000	+0.010	+0.019	+0.024	+0.032	+0.037
TO	10	−0.170	−0.076	−0.028	−0.014	−0.090	−0.036	−0.015	−0.009	+0.001	+0.010	+0.015	+0.023	+0.028
OVER	10	−0.095	−0.050	−0.016	−0.006	0.000	0.000	0.000	0.000	+0.012	+0.023	+0.029	+0.039	+0.044
TO	14	−0.205	−0.093	−0.034	−0.017	−0.110	−0.043	−0.018	−0.011	+0.001	+0.012	+0.018	+0.028	+0.033
OVER	14	−0.095	−0.050	−0.016	−0.006	0.000	0.000	0.000	0.000	+0.012	+0.023	+0.029	+0.039	+0.044
TO	18	−0.205	−0.093	−0.034	−0.017	−0.110	−0.043	−0.018	−0.011	+0.001	+0.012	+0.018	+0.028	+0.033
OVER	18	−0.110	−0.065	−0.020	−0.007	0.000	0.000	0.000	0.000	+0.015	+0.028	+0.035	+0.048	+0.054
TO	24	−0.240	−0.117	−0.041	−0.020	−0.130	−0.052	−0.021	−0.013	+0.002	+0.015	+0.022	+0.035	+0.041
OVER	24	−0.010	−0.065	−0.020	−0.007	0.000	0.000	0.000	0.000	+0.015	+0.028	+0.035	+0.048	+0.061
TO	30	−0.240	−0.117	−0.041	−0.020	−0.130	−0.052	−0.021	−0.013	+0.002	+0.015	+0.022	+0.035	+0.048
OVER	30	−0.120	−0.080	−0.025	−0.009	0.000	0.000	0.000	0.000	+0.018	+0.033	+0.042	+0.059	+0.076
TO	40	−0.280	−0.142	−0.050	−0.025	−0.160	−0.062	−0.025	−0.016	+0.002	+0.017	+0.026	+0.043	+0.060
OVER	40	−0.130	−0.080	−0.025	−0.009	0.000	0.000	0.000	0.000	+0.018	+0.033	+0.042	+0.059	+0.086
TO	50	−0.290	−0.142	−0.050	−0.025	−0.160	−0.062	−0.025	−0.016	+0.002	+0.017	+0.026	+0.043	+0.070
OVER	50	−0.140	−0.100	−0.030	−0.010	0.000	0.000	0.000	0.000	+0.021	+0.039	+0.051	+0.072	+0.106
TO	65	−0.330	−0.174	−0.060	−0.029	−0.190	−0.074	−0.030	−0.019	+0.002	+0.020	+0.032	+0.053	+0.087
OVER	65	−0.150	−0.100	−0.030	−0.010	0.000	0.000	0.000	0.000	+0.021	+0.039	+0.051	+0.078	+0.121
TO	80	−0.340	−0.174	−0.060	−0.029	−0.190	−0.074	−0.030	−0.019	+0.002	+0.020	+0.032	+0.059	+0.102
OVER	80	−0.170	−0.120	−0.036	−0.012	0.000	0.000	0.000	0.000	+0.025	+0.045	+0.059	+0.093	+0.146
TO	100	−0.390	−0.207	−0.071	−0.034	−0.220	−0.087	−0.035	−0.022	+0.003	+0.023	+0.037	+0.071	+0.124
OVER	100	−0.180	−0.120	−0.036	−0.012	0.000	0.000	0.000	0.000	+0.025	+0.045	+0.059	+0.101	+0.166
TO	120	−0.400	−0.207	−0.071	−0.034	−0.220	−0.087	−0.035	−0.022	+0.003	+0.023	+0.037	+0.079	+0.144
OVER	120	−0.200	−0.145	−0.043	−0.014	0.000	0.000	0.000	0.000	+0.028	+0.052	+0.068	+0.117	+0.195
TO	140	−0.450	−0.245	−0.083	−0.039	−0.250	−0.100	−0.040	−0.025	+0.003	+0.027	+0.043	+0.092	+0.170
OVER	140	−0.210	−0.145	−0.043	−0.014	0.000	0.000	0.000	0.000	+0.028	+0.052	+0.068	+0.125	+0.215
TO	160	−0.460	−0.245	−0.083	−0.039	−0.250	−0.100	−0.040	−0.025	+0.003	+0.027	+0.043	+0.100	+0.190
OVER	160	−0.230	−0.145	−0.043	−0.014	0.000	0.000	0.000	0.000	+0.028	+0.052	+0.068	+0.133	+0.235
TO	180	−0.480	−0.245	−0.083	−0.039	−0.250	−0.100	−0.040	−0.025	+0.003	+0.027	+0.043	+0.108	+0.210
OVER	180	−0.240	−0.170	−0.050	−0.015	0.000	0.000	0.000	0.000	+0.033	+0.060	+0.079	+0.151	+0.265
TO	200	−0.530	−0.285	−0.096	−0.044	−0.290	−0.115	−0.046	−0.029	+0.004	+0.031	+0.050	+0.122	+0.236
OVER	200	−0.260	−0.170	−0.050	−0.015	0.000	0.000	0.000	0.000	+0.033	+0.060	+0.079	+0.159	+0.287
TO	225	−0.550	−0.285	−0.096	−0.044	−0.290	−0.115	−0.046	−0.029	+0.004	+0.031	+0.050	+0.130	+0.258
OVER	225	−0.280	−0.170	−0.050	−0.015	0.000	0.000	0.000	0.000	+0.033	+0.060	+0.079	+0.169	+0.313
TO	250	−0.570	−0.285	−0.096	−0.044	−0.290	−0.115	−0.046	−0.029	+0.004	+0.031	+0.050	+0.140	+0.284
OVER	250	−0.300	−0.190	−0.056	−0.017	0.000	0.000	0.000	0.000	+0.036	+0.066	+0.088	+0.190	+0.347
TO	280	−0.620	−0.320	−0.108	−0.049	−0.320	−0.130	−0.052	−0.032	+0.004	+0.034	+0.056	+0.158	+0.315
OVER	280	−0.330	−0.190	−0.056	−0.017	−0.000	0.000	0.000	0.000	+0.036	+0.066	+0.088	+0.202	+0.382
TO	315	−0.650	−0.320	−0.108	−0.049	−0.320	−0.130	−0.052	−0.032	+0.004	+0.034	+0.056	+0.170	+0.350
OVER	315	−0.360	−0.210	−0.062	−0.018	0.000	0.000	0.000	0.000	+0.040	+0.073	+0.098	+0.226	+0.426
TO	355	−0.720	−0.350	−0.119	−0.054	−0.360	−0.140	−0.057	−0.036	+0.004	+0.037	+0.062	+0.190	+0.390
OVER	355	−0.400	−0.210	−0.062	−0.018	0.000	0.000	0.000	0.000	+0.040	+0.073	+0.098	+0.244	+0.471
TO	400	−0.760	−0.350	−0.119	−0.054	−0.360	−0.140	−0.057	−0.036	+0.004	+0.037	+0.062	+0.208	+0.435
OVER	400	−0.440	−0.230	−0.068	−0.020	0.000	0.000	0.000	0.000	+0.045	+0.080	+0.108	+0.272	+0.530
TO	450	−0.840	−0.385	−0.131	−0.060	−0.400	−0.155	−0.063	−0.040	+0.005	+0.040	+0.068	+0.232	+0.490
OVER	450	−0.480	−0.230	−0.068	−0.020	0.000	0.000	0.000	0.000	+0.045	+0.080	+0.108	+0.292	+0.580
TO	500	−0.880	−0.385	−0.131	−0.060	−0.400	−0.155	−0.063	−0.040	+0.005	+0.040	+0.068	+0.252	+0.540

Table 6-3 Preferred Hole Basis Clearance Fits

Dimensions in Millimeters

	HOLE H11	SHAFT c11	DIFF	HOLE H9	SHAFT d9	DIFF	HOLE H8	SHAFT f7	DIFF	HOLE H7	SHAFT g6	DIFF	HOLE H7	SHAFT h6	DIFF
MAX	1.060	0.940	0.180	1.025	0.980	0.070	1.014	0.994	0.030	1.010	0.998	0.018	1.010	1.000	0.016
MIN	1.000	0.880	0.060	1.000	0.955	0.020	1.000	0.984	0.006	1.000	0.992	0.002	1.000	0.994	0.000
MAX	1.260	1.140	0.180	1.225	1.180	0.070	1.214	1.194	0.030	1.210	1.198	0.018	1.210	1.200	0.016
MIN	1.200	1.080	0.060	1.200	1.155	0.020	1.200	1.184	0.006	1.200	1.192	0.002	1.200	1.194	0.000
MAX	1.660	1.540	0.180	1.625	1.580	0.070	1.614	1.594	0.030	1.610	1.598	0.018	1.610	1.600	0.016
MIN	1.600	1.480	0.060	1.600	1.555	0.020	1.600	1.584	0.006	1.600	1.592	0.002	1.600	1.594	0.000
MAX	2.060	1.940	0.180	2.025	1.980	0.070	2.014	1.994	0.030	2.010	1.998	0.018	2.010	2.000	0.016
MIN	2.000	1.880	0.060	2.000	1.955	0.020	2.000	1.984	0.006	2.000	1.992	0.002	2.000	1.994	0.000
MAX	2.560	2.440	0.180	2.525	2.480	0.070	2.514	2.494	0.030	2.510	2.498	0.018	2.510	2.500	0.016
MIN	2.500	2.380	0.060	2.500	2.455	0.020	2.500	2.484	0.006	2.500	2.492	0.002	2.500	2.494	0.000
MAX	3.060	2.940	0.180	3.025	2.980	0.070	3.014	2.994	0.030	3.010	2.998	0.018	3.010	3.000	0.016
MIN	3.000	2.880	0.060	3.000	2.955	0.020	3.000	2.984	0.006	3.000	2.992	0.002	3.000	2.994	0.000
MAX	4.075	3.930	0.220	4.030	3.970	0.090	4.018	3.990	0.040	4.012	3.996	0.024	4.012	4.000	0.020
MIN	4.000	3.855	0.070	4.000	3.940	0.030	4.000	3.978	0.010	4.000	3.988	0.004	4.000	3.992	0.000
MAX	5.075	4.930	0.220	5.030	4.970	0.090	5.018	4.990	0.040	5.012	4.996	0.024	5.012	5.000	0.020
MIN	5.000	4.855	0.070	5.000	4.949	0.030	5.000	4.978	0.010	5.000	4.988	0.004	5.000	4.992	0.000
MAX	6.075	5.930	0.220	6.030	5.970	0.090	6.018	5.990	0.040	6.012	5.996	0.024	6.012	6.000	0.020
MIN	6.000	5.855	0.070	6.000	5.940	0.030	6.000	5.978	0.010	6.000	5.988	0.004	6.000	5.992	0.000
MAX	8.090	7.920	0.260	8.036	7.960	0.012	8.022	7.987	0.050	8.015	7.995	0.029	8.015	8.000	0.024
MIN	8.000	7.830	0.080	8.000	7.924	0.040	8.000	7.972	0.013	8.000	7.986	0.005	8.000	7.991	0.000
MAX	10.090	9.920	0.260	10.036	9.960	0.112	10.022	9.987	0.050	10.015	9.995	0.029	10.015	10.000	0.024
MIN	10.000	9.830	0.080	10.000	9.924	0.040	10.000	9.972	0.013	10.000	9.286	0.005	10.000	9.991	0.000
MAX	12.110	11.905	0.315	12.043	11.950	0.136	12.027	11.984	0.061	12.018	11.994	0.035	12.018	12.000	0.029
MIN	12.000	11.795	0.095	12.000	11.907	0.050	12.000	11.966	0.016	12.000	11.983	0.006	12.000	11.989	0.000
MAX	16.110	15.905	0.315	16.043	15.950	0.136	16.027	15.984	0.061	16.018	15.994	0.035	16.018	16.000	0.029
MIN	16.000	15.795	0.095	16.000	15.907	0.050	16.000	15.966	0.016	16.000	15.983	0.006	16.000	15.989	0.000
MAX	20.130	19.890	0.370	20.052	19.935	0.169	20.033	19.980	0.074	20.021	19.993	0.041	20.021	20.000	0.034
MIN	20.000	19.760	0.110	20.000	19.883	0.065	20.000	19.959	0.020	20.000	19.980	0.007	20.000	19.987	0.000
MAX	25.130	24.890	0.370	25.052	24.935	0.169	25.033	24.980	0.074	25.021	24.993	0.041	25.021	25.000	0.034
MIN	25.000	24.760	0.110	25.000	24.883	0.065	25.000	24.959	0.020	25.000	24.980	0.007	25.000	24.987	0.000
MAX	30.130	29.890	0.370	30.052	29.935	0.169	30.033	29.980	0.074	30.021	29.993	0.041	30.021	30.000	0.034
MIN	30.000	29.760	0.110	30.000	29.883	0.065	30.000	29.959	0.020	30.000	29.980	0.007	30.000	29.987	0.000
MAX	40.160	39.880	0.440	40.062	39.920	0.204	40.039	39.975	0.089	40.025	39.991	0.050	40.025	40.000	0.041
MIN	40.000	39.720	0.120	40.000	39.858	0.080	40.000	39.950	0.025	40.000	39.975	0.009	40.000	39.984	0.000
MAX	50.160	49.870	0.450	50.062	49.920	0.204	50.039	49.975	0.089	50.025	49.991	0.050	50.025	50.000	0.041
MIN	50.000	49.710	0.130	50.000	49.858	0.080	50.000	49.950	0.025	50.000	49.975	0.009	50.000	49.984	0.000
MAX	60.190	59.860	0.520	60.074	59.900	0.248	60.046	59.970	0.106	60.030	59.990	0.059	60.030	60.000	0.049
MIN	60.000	59.670	0.140	60.000	59.826	0.100	60.000	59.940	0.030	60.000	59.971	0.010	60.000	59.981	0.000
MAX	80.190	79.850	0.530	80.074	79.900	0.248	80.046	79.970	0.106	80.030	79.990	0.059	80.030	80.000	0.049
MIN	80.000	79.660	0.150	80.000	79.826	0.100	80.000	79.940	0.030	80.000	79.971	0.010	80.000	79.981	0.000
MAX	100.220	99.830	0.610	100.087	99.880	0.294	100.054	99.964	0.125	100.035	99.988	0.069	100.035	100.000	0.057
MIN	100.000	99.610	0.170	100.000	99.793	0.120	100.000	99.929	0.036	100.000	99.966	0.012	100.000	99.978	0.000
MAX	120.220	119.820	0.620	120.087	119.880	0.294	120.054	119.964	0.125	120.035	119.988	0.069	120.035	120.000	0.057
MIN	120.000	119.600	0.180	120.000	119.793	0.120	120.000	119.929	0.036	120.000	119.966	0.012	120.000	119.978	0.000
MAX	160.250	159.790	0.710	160.100	159.855	0.345	160.063	159.957	0.146	160.040	159.986	0.079	160.040	160.000	0.065
MIN	160.000	159.540	0.210	160.000	159.755	0.145	160.000	159.917	0.043	160.000	159.961	0.014	160.000	159.975	0.000
MAX	200.290	199.760	0.820	200.115	199.830	0.400	200.072	199.950	0.168	200.046	199.985	0.090	200.046	200.000	0.075
MIN	200.000	199.470	0.240	200.000	199.715	0.170	200.000	199.904	0.050	200.000	199.956	0.015	200.000	199.971	0.000
MAX	250.290	249.720	0.860	250.115	249.830	0.400	250.072	249.950	0.168	250.046	249.985	0.090	250.046	250.000	0.075
MIN	250.000	249.430	0.280	250.000	249.715	0.170	250.000	249.904	0.050	250.000	249.956	0.015	250.000	249.971	0.000
MAX	300.320	299.670	0.970	300.130	299.810	0.450	300.081	299.944	0.189	300.052	299.983	0.101	300.052	300.000	0.084
MIN	300.000	299.350	0.330	300.000	299.680	0.190	300.000	299.892	0.056	300.000	299.951	0.017	300.000	299.968	0.000
MAX	400.360	399.600	1.120	400.140	399.790	0.490	400.089	399.938	0.208	400.057	399.982	0.111	400.057	400.000	0.093
MIN	400.000	399.240	0.400	400.000	399.650	0.210	400.000	399.881	0.062	400.000	399.946	0.018	400.000	399.964	0.000
MAX	500.400	499.520	1.280	500.155	499.770	0.540	500.097	499.932	0.228	500.063	499.980	0.123	500.063	500.000	0.103
MIN	500.000	499.120	0.480	500.000	499.615	0.230	500.000	499.869	0.068	500.000	499.940	0.020	500.000	499.960	0.000

Table 6–4   Preferred Hole Basis Transition and Interference Fits

Dimensions in Millimeters

	HOLE H7	SHAFT k6	DIFF	HOLE H7	SHAFT n6	DIFF	HOLE H7	SHAFT p6	DIFF	HOLE H7	SHAFT s6	DIFF	HOLE H7	SHAFT u6	DIFF
MAX	1.010	1.006	0.010	1.010	1.010	0.006	1.010	1.012	0.004	1.010	1.020	−0.004	1.010	1.024	−0.008
MIN	1.000	1.000	−0.006	1.000	1.004	−0.010	1.000	1.006	−0.012	1.000	1.014	−0.020	1.000	1.018	−0.024
MAX	1.210	1.206	0.010	1.210	1.210	0.006	1.210	1.212	0.004	1.210	1.220	−0.004	1.210	1.224	−0.008
MIN	1.200	1.200	−0.006	1.200	1.204	−0.010	1.200	1.206	−0.012	1.200	1.214	−0.020	1.200	1.218	−0.024
MAX	1.610	1.606	0.010	1.610	1.610	0.006	1.610	1.612	0.004	1.610	1.620	−0.004	1.610	1.624	−0.008
MIN	1.600	1.600	−0.006	1.600	1.604	−0.010	1.600	1.606	−0.012	1.600	1.614	−0.020	1.600	1.618	−0.024
MAX	2.010	2.006	0.010	2.010	2.010	0.006	2.010	2.012	0.004	2.010	2.020	−0.004	2.010	2.024	−0.008
MIN	2.000	2.000	−0.006	2.000	2.004	−0.010	2.000	2.006	−0.012	2.000	2.014	−0.080	2.000	2.018	−0.024
MAX	2.510	2.506	0.010	2.510	2.510	0.006	2.510	2.512	0.004	2.510	2.520	−0.004	2.510	2.524	−0.008
MIN	2.500	2.500	−0.006	2.500	2.504	−0.010	2.500	2.506	−0.012	2.500	2.514	−0.020	2.500	2.518	−0.024
MAX	3.010	3.006	0.010	3.010	3.010	0.006	3.010	3.012	0.004	3.010	3.020	−0.004	3.010	3.024	−0.008
MIN	3.000	3.000	−0.006	3.000	3.004	−0.010	3.000	3.006	−0.012	3.000	3.014	−0.020	3.000	3.018	−0.024
MAX	4.012	4.009	0.011	4.012	4.016	0.004	4.012	4.020	0.000	4.012	4.027	−0.007	4.012	4.031	−0.011
MIN	4.000	4.001	−0.009	4.000	4.008	−0.016	4.000	4.012	−0.020	4.000	4.019	−0.027	4.000	4.023	−0.031
MAX	5.012	5.009	0.011	5.012	5.016	0.004	5.012	5.020	0.000	5.012	5.027	−0.007	5.012	5.031	−0.011
MIN	5.000	5.001	−0.009	5.000	5.008	−0.016	5.000	5.012	−0.020	5.000	5.019	−0.027	5.000	5.023	−0.031
MAX	6.012	6.009	0.011	6.012	6.016	0.004	6.012	6.020	0.000	6.012	6.027	−0.007	6.012	6.031	−0.011
MIN	6.000	6.001	−0.009	6.000	6.008	−0.016	6.000	6.012	−0.020	6.000	6.019	−0.027	6.000	6.023	−0.031
MAX	8.015	8.010	0.014	8.015	8.019	0.005	8.015	8.024	0.000	8.015	8.032	−0.008	8.015	8.037	−0.013
MIN	8.000	8.001	−0.010	8.000	8.010	−0.019	8.000	8.015	−0.024	8.000	8.023	−0.032	8.000	8.028	−0.037
MAX	10.015	10.010	0.014	10.015	10.019	0.005	10.015	10.024	0.000	10.015	10.032	−0.008	10.015	10.037	−0.013
MIN	10.000	10.001	−0.010	10.000	10.010	−0.019	10.000	10.015	−0.024	10.000	10.023	−0.032	10.000	10.028	−0.037
MAX	12.018	12.012	0.017	12.018	12.023	0.006	12.018	12.029	0.000	12.018	12.039	−0.010	12.018	12.044	−0.015
MIN	12.000	12.001	−0.012	12.000	12.012	−0.023	12.000	12.018	−0.029	12.000	12.028	−0.039	12.000	12.033	−0.044
MAX	16.018	16.012	0.017	16.018	16.023	0.006	16.018	16.029	0.000	16.018	16.039	−0.010	16.018	16.044	−0.015
MIN	16.000	16.001	−0.012	16.000	16.012	−0.023	16.000	16.018	−0.029	16.000	16.028	−0.039	16.000	16.033	−0.044
MAX	20.021	20.015	0.019	20.021	20.028	0.006	20.021	20.035	−0.001	20.021	20.048	−0.014	20.021	20.054	−0.020
MIN	20.000	20.002	−0.015	20.000	20.015	−0.028	20.000	20.022	−0.035	20.000	20.035	−0.048	20.000	20.041	−0.054
MAX	25.021	25.015	0.019	25.021	25.028	0.006	25.021	25.035	−0.001	25.021	25.048	−0.014	25.021	25.061	−0.027
MIN	25.000	25.002	−0.015	25.000	25.015	−0.028	25.000	25.022	−0.035	25.000	25.035	−0.048	25.000	25.048	−0.061
MAX	30.021	30.015	0.019	30.021	30.028	0.006	30.021	30.035	−0.001	30.021	30.048	−0.014	30.021	30.061	−0.027
MIN	30.000	30.002	−0.015	30.000	30.015	−0.028	30.000	30.022	−0.035	30.000	30.035	−0.048	30.000	30.048	−0.061
MAX	40.025	40.018	0.023	40.025	40.033	0.008	40.025	40.042	−0.001	40.025	40.059	−0.018	40.025	40.076	−0.035
MIN	40.000	40.002	−0.018	40.000	40.017	−0.033	40.000	40.026	−0.042	40.000	40.043	−0.059	40.000	40.060	−0.076
MAX	50.025	50.018	0.023	50.025	50.033	0.008	50.025	50.042	−0.001	50.025	50.059	−0.018	50.025	50.086	−0.045
MIN	50.000	50.002	−0.018	50.000	50.017	−0.033	50.000	50.026	−0.042	50.000	50.043	−0.059	50.000	50.070	−0.086
MAX	60.030	60.021	−0.028	60.030	60.039	0.010	60.030	60.051	−0.002	60.030	60.072	−0.023	60.030	60.106	−0.057
MIN	60.000	60.002	−0.021	60.000	60.020	−0.039	60.000	60.032	−0.051	60.000	60.053	−0.072	60.000	60.087	−0.106
MAX	80.030	80.021	0.028	80.030	80.039	0.010	80.030	80.051	−0.002	80.030	80.078	−0.029	80.030	80.121	−0.072
MIN	80.000	80.002	−0.021	80.000	80.020	−0.039	80.000	80.032	−0.051	80.000	80.059	−0.078	80.000	80.102	−0.121
MAX	100.035	100.025	0.032	100.035	100.045	0.012	100.035	100.059	−0.002	100.035	100.093	−0.036	100.035	100.146	−0.089
MIN	100.000	100.003	−0.025	100.000	100.023	−0.045	100.000	100.037	−0.059	100.000	100.071	−0.093	100.000	100.124	−0.146
MAX	120.035	120.025	0.032	120.035	120.045	0.012	120.035	120.059	−0.002	120.035	120.101	−0.044	120.035	120.166	−0.109
MIN	120.000	120.003	−0.025	120.000	120.023	−0.045	120.000	120.037	−0.059	120.000	120.079	−0.101	120.000	120.144	−0.166
MAX	160.040	160.028	0.037	160.040	160.052	0.013	160.040	160.068	−0.003	160.040	160.125	−0.060	160.040	160.215	−0.150
MIN	160.000	160.003	−0.028	160.000	160.027	−0.052	160.000	160.043	−0.068	160.000	160.100	−0.125	160.000	160.190	−0.215
MAX	200.046	200.033	0.042	200.046	200.060	0.015	200.046	200.079	−0.004	200.046	200.151	−0.076	200.046	200.265	−0.190
MIN	200.000	200.004	−0.033	200.000	200.031	−0.060	200.000	200.050	−0.079	200.000	200.122	−0.151	200.000	200.236	−0.265
MAX	250.046	250.033	0.042	250.046	250.060	0.015	250.046	250.079	−0.004	250.046	250.169	−0.094	250.046	250.313	−0.238
MIN	250.000	250.004	−0.033	250.000	250.031	−0.060	250.000	250.050	−0.079	250.000	250.140	−0.169	250.000	250.284	−0.313
MAX	300.052	300.036	0.048	300.052	300.066	0.018	300.052	300.088	−0.004	300.052	300.202	−0.118	300.052	300.382	−0.298
MIN	300.000	300.004	−0.036	300.000	300.034	−0.066	300.000	300.056	−0.088	300.000	300.170	−0.202	300.000	300.350	−0.382
MAX	400.057	400.040	0.053	400.057	400.073	0.020	400.057	400.098	−0.005	400.057	400.244	−0.151	400.057	400.471	−0.378
MIN	400.000	400.004	−0.040	400.000	400.037	−0.073	400.000	400.062	−0.098	400.000	400.208	−0.244	400.000	400.435	−0.471
MAX	500.063	500.045	0.058	500.063	500.080	0.023	500.063	500.108	−0.005	500.063	500.292	−0.189	500.063	500.580	−0.477
MIN	500.000	500.005	−0.045	500.000	500.040	−0.080	500.000	500.068	−0.108	500.000	500.252	−0.292	500.000	500.540	−0.580

Table 6–5 Preferred Shaft Basis Clearance Fits

Dimensions in Millimeters

	HOLE C11	SHAFT h11	DIFF	HOLE D9	SHAFT h9	DIFF	HOLE F8	SHAFT h7	DIFF	HOLE G7	SHAFT h6	DIFF	HOLE H7	SHAFT h6	DIFF
MAX	1.120	1.000	0.180	1.045	1.000	0.070	1.020	1.000	0.030	1.012	1.000	0.018	1.010	1.000	0.016
MIN	1.060	0.940	0.060	1.020	0.975	0.020	1.006	0.990	0.006	1.002	0.994	0.002	1.000	0.994	0.000
MAX	1.320	1.200	0.180	1.245	1.200	0.070	1.220	1.200	0.030	1.212	1.200	0.018	1.210	1.200	0.016
MIN	1.260	1.140	0.060	1.220	1.175	0.020	1.206	1.190	0.006	1.202	1.194	0.002	1.200	1.194	0.000
MAX	1.720	1.600	0.180	1.645	1.600	0.070	1.620	1.600	0.030	1.612	1.600	0.018	1.610	1.600	0.016
MIN	1.660	1.540	0.060	1.620	1.575	0.020	1.606	1.550	0.006	1.602	1.594	0.002	1.600	1.594	0.000
MAX	2.120	2.000	0.180	2.045	2.000	0.070	2.020	2.000	0.030	2.012	2.000	0.018	2.010	2.000	0.016
MIN	2.060	1.940	0.060	2.020	1.975	0.020	2.006	1.990	0.006	2.002	1.994	0.002	2.000	1.994	0.000
MAX	2.620	2.500	0.180	2.545	2.500	0.070	2.520	2.500	0.030	2.512	2.500	0.018	2.510	2.500	0.016
MIN	2.560	2.440	0.060	2.520	2.475	0.020	2.506	2.490	0.006	2.502	2.494	0.002	2.500	2.494	0.000
MAX	3.120	3.000	0.180	3.045	3.000	0.070	3.020	3.000	0.030	3.012	3.000	0.018	3.010	3.000	0.016
MIN	3.060	2.940	0.060	3.020	2.975	0.020	3.006	2.990	0.006	3.002	2.994	0.002	3.000	2.994	0.000
MAX	4.145	4.000	0.220	4.060	4.000	0.090	4.028	4.000	0.040	4.016	4.000	0.024	4.012	4.000	0.020
MIN	4.070	3.925	0.070	4.030	3.970	0.030	4.010	3.988	0.010	4.004	3.992	0.004	4.000	3.992	0.000
MAX	5.145	5.000	0.220	5.060	5.000	0.090	5.028	5.000	0.040	5.016	5.000	0.024	5.012	5.000	0.020
MIN	5.070	4.925	0.070	5.030	4.970	0.030	5.010	4.988	0.010	5.004	4.992	0.004	5.000	4.992	0.000
MAX	6.145	6.000	0.220	6.060	6.000	0.090	6.028	6.000	0.040	6.016	6.000	0.024	6.012	6.000	0.020
MIN	6.070	5.925	0.070	6.030	5.970	0.030	6.010	5.988	0.010	6.004	5.992	0.004	6.000	5.992	0.000
MAX	8.170	8.000	0.260	8.076	8.000	0.112	8.035	8.000	0.050	8.020	8.000	0.029	8.015	8.000	0.024
MIN	8.080	7.910	0.080	8.040	7.964	0.040	8.013	7.985	0.013	8.005	7.991	0.005	8.000	7.991	0.000
MAX	10.170	10.000	0.260	10.076	10.000	0.112	10.035	10.000	0.050	10.020	10.000	0.029	10.015	10.000	0.024
MIN	10.080	9.910	0.080	10.040	9.964	0.040	10.013	9.985	0.013	10.005	9.991	0.005	10.000	9.991	0.000
MAX	12.205	12.000	0.315	12.093	12.000	0.136	12.043	12.000	0.061	12.024	12.000	0.035	12.018	12.000	0.029
MIN	12.095	11.890	0.095	12.050	11.957	0.050	12.016	11.982	0.016	12.006	11.989	0.006	12.000	11.989	0.000
MAX	16.205	16.000	0.315	16.093	16.000	0.136	16.043	16.000	0.061	16.024	16.000	0.035	16.018	16.000	0.029
MIN	16.095	15.890	0.095	16.050	15.957	0.050	16.016	15.982	0.016	16.006	15.989	0.006	16.000	15.989	0.000
MAX	20.240	20.000	0.370	20.117	20.000	0.169	20.053	20.000	0.074	20.028	20.000	0.041	20.021	20.000	0.034
MIN	20.110	19.870	0.110	20.065	19.948	0.065	20.020	19.979	0.020	20.007	19.987	0.007	20.000	19.987	0.000
MAX	25.240	25.000	0.370	25.117	25.000	0.169	25.053	25.000	0.074	25.028	25.000	0.041	25.021	25.000	0.034
MIN	25.110	24.870	0.110	25.065	24.948	0.065	25.020	24.979	0.020	25.007	24.987	0.007	25.000	24.987	0.000
MAX	30.240	30.000	0.370	30.117	30.000	0.169	30.053	30.000	0.074	30.028	30.000	0.041	30.021	30.000	0.034
MIN	30.110	29.870	0.110	30.065	29.948	0.065	30.020	29.979	0.020	30.007	29.987	0.007	30.000	29.987	0.000
MAX	40.280	40.000	0.440	40.142	40.000	0.204	40.064	40.000	0.089	40.034	40.000	0.050	40.025	40.000	0.041
MIN	40.120	39.840	0.120	40.080	39.938	0.080	40.025	39.975	0.025	40.009	39.984	0.009	40.000	39.984	0.000
MAX	50.290	50.000	0.450	50.142	50.000	0.204	50.064	50.000	0.089	50.034	50.000	0.050	50.025	50.000	0.041
MIN	50.130	49.840	0.130	50.080	49.938	0.080	50.025	49.975	0.025	50.009	49.984	0.009	50.000	49.984	0.000
MAX	60.330	60.000	0.520	60.174	60.000	0.248	60.076	60.000	0.106	60.040	60.000	0.059	60.030	60.000	0.049
MIN	60.140	59.810	0.140	60.100	59.926	0.100	60.030	59.970	0.030	60.010	59.981	0.010	60.000	59.981	0.000
MAX	80.340	80.000	0.530	80.174	80.000	0.248	80.076	80.000	0.106	80.040	80.000	0.059	80.030	80.000	0.049
MIN	80.150	79.810	0.150	80.100	79.926	0.100	80.030	79.970	0.030	80.010	79.981	0.010	80.000	79.981	0.000
MAX	100.390	100.000	0.610	100.207	100.000	0.294	100.090	100.000	0.125	100.047	100.000	0.069	100.035	100.000	0.057
MIN	100.170	99.780	0.170	100.120	99.913	0.120	100.036	99.965	0.036	100.012	99.978	0.012	100.000	99.978	0.000
MAX	120.400	120.000	0.620	120.207	120.000	0.294	120.090	120.000	0.125	120.047	120.000	0.069	120.035	120.000	0.057
MIN	120.180	119.780	0.180	120.120	119.913	0.120	120.036	119.965	0.036	120.012	119.978	0.012	120.000	119.978	0.000
MAX	160.460	160.000	0.710	160.245	160.000	0.345	160.106	160.000	0.146	160.054	160.000	0.079	160.040	160.000	0.065
MIN	160.210	159.750	0.210	160.145	159.900	0.145	160.043	159.960	0.043	160.014	159.975	0.014	160.000	159.975	0.000
MAX	200.530	200.000	0.820	200.285	200.000	0.400	200.122	200.000	0.168	200.061	200.000	0.090	200.046	200.000	0.075
MIN	200.240	199.710	0.240	200.170	199.885	0.170	200.050	199.954	0.050	200.015	199.971	0.015	200.000	199.971	0.000
MAX	250.570	250.000	0.860	250.285	250.000	0.400	250.122	250.000	0.168	250.061	250.000	0.090	250.046	250.000	0.075
MIN	250.280	249.710	0.280	250.170	249.885	0.170	250.050	249.954	0.050	250.015	249.971	0.015	250.000	249.971	0.000
MAX	300.650	300.000	0.970	300.320	300.000	0.450	300.137	300.000	0.189	300.069	300.000	0.101	300.052	300.000	0.084
MIN	300.330	299.680	0.330	300.190	299.870	0.190	300.056	299.948	0.056	300.017	255.968	0.017	300.000	299.968	0.000
MAX	400.760	400.000	1.120	400.350	400.000	0.490	400.151	400.000	0.208	400.075	400.000	0.111	400.057	400.000	0.093
MIN	400.400	399.640	0.400	400.210	399.860	0.210	400.062	399.943	0.062	400.018	399.964	0.018	400.000	399.964	0.000
MAX	500.880	500.000	1.280	500.385	500.000	0.540	500.165	500.000	0.228	500.083	500.000	0.123	500.063	500.000	0.103
MIN	500.480	499.600	0.480	500.230	499.845	0.230	500.068	499.937	0.068	500.020	499.960	0.020	500.000	499.960	0.000

Table 6-6 Preferred Shaft Basis Transition and Interference Fits

Dimensions in Millimeters

	HOLE K7	SHAFT h6	DIFF	HOLE N7	SHAFT h6	DIFF	HOLE P7	SHAFT h6	DIFF	HOLE S7	SHAFT h6	DIFF	HOLE U7	SHAFT h6	DIFF
MAX	1.000	1.000	0.006	0.996	1.000	0.002	0.994	1.000	0.000	0.986	1.000	−0.008	0.982	1.000	−0.012
MIN	0.990	0.994	−0.010	0.986	0.994	−0.014	0.984	0.994	−0.016	0.976	0.994	−0.024	0.972	0.994	−0.028
MAX	1.200	1.200	0.006	1.196	1.200	0.002	1.194	1.200	0.000	1.186	1.200	−0.008	1.182	1.200	−0.012
MIN	1.190	1.194	−0.010	1.186	1.194	−0.014	1.184	1.194	−0.016	1.176	1.194	−0.024	1.172	1.194	−0.028
MAX	1.600	1.600	0.006	1.596	1.600	0.002	1.594	1.600	0.000	1.586	1.600	−0.008	1.582	1.600	−0.012
MIN	1.590	1.594	−0.010	1.586	1.594	−0.014	1.584	1.594	−0.016	1.576	1.594	−0.024	1.572	1.594	−0.028
MAX	2.000	2.000	0.006	1.996	2.000	0.002	1.994	2.000	0.000	1.986	2.000	−0.008	1.982	2.000	−0.012
MIN	1.990	1.994	−0.010	1.986	1.994	−0.014	1.984	1.994	−0.016	1.976	1.994	−0.024	1.972	1.994	−0.028
MAX	2.500	2.500	0.006	2.496	2.500	0.002	2.494	2.500	0.000	2.486	2.500	−0.008	2.482	2.500	−0.012
MIN	2.490	2.494	−0.010	2.486	2.494	−0.014	2.484	2.494	−0.016	2.476	2.494	−0.024	2.472	2.494	−0.028
MAX	3.000	3.000	0.006	2.996	3.000	0.002	2.994	3.000	0.000	2.986	3.000	−0.008	2.982	3.000	−0.012
MIN	2.990	2.994	−0.010	2.986	2.994	−0.014	2.984	2.994	−0.016	2.976	2.994	−0.024	2.972	2.994	−0.028
MAX	4.003	4.000	0.011	3.996	4.000	0.004	3.992	4.000	0.000	3.985	4.000	−0.007	3.981	4.000	−0.011
MIN	3.991	3.992	−0.009	3.984	3.992	−0.016	3.980	3.992	−0.020	3.973	3.992	−0.027	3.969	3.992	−0.031
MAX	5.003	5.000	0.011	4.996	5.000	0.004	4.992	5.000	0.000	4.985	5.000	−0.007	4.981	5.000	−0.011
MIN	4.991	4.992	−0.009	4.984	4.992	−0.016	4.980	4.992	−0.020	4.973	4.992	−0.027	4.969	4.992	−0.031
MAX	6.003	6.000	0.011	5.996	6.000	0.004	5.992	6.000	0.000	5.985	6.000	−0.007	5.981	6.000	−0.011
MIN	5.991	5.992	−0.009	5.984	5.992	−0.016	5.980	5.992	−0.020	5.973	5.992	−0.027	5.969	5.992	−0.031
MAX	8.005	8.000	0.014	7.996	8.000	0.005	7.991	8.000	0.000	7.983	8.000	−0.008	7.978	8.000	−0.013
MIN	7.990	7.991	−0.010	7.981	7.991	−0.019	7.976	7.991	−0.024	7.968	7.991	−0.032	7.963	7.991	−0.037
MAX	10.005	10.000	0.014	9.996	10.000	0.005	9.991	10.000	0.000	9.983	10.000	−0.008	9.978	10.000	−0.013
MIN	9.990	9.991	−0.010	9.981	9.991	−0.019	9.976	9.991	−0.024	9.968	9.991	−0.032	9.963	9.991	−0.037
MAX	12.006	12.000	0.017	11.995	12.000	0.006	11.989	12.000	0.000	11.979	12.000	−0.010	11.974	12.000	−0.015
MIN	11.988	11.989	−0.012	11.977	11.989	−0.023	11.971	11.989	−0.029	11.961	11.989	−0.039	11.956	11.989	−0.044
MAX	16.006	16.000	0.017	15.995	16.000	0.006	15.989	16.000	0.000	15.979	16.000	−0.010	15.974	16.000	−0.015
MIN	15.988	15.989	−0.012	15.977	15.989	−0.023	15.971	15.989	−0.029	15.961	15.989	−0.039	15.956	15.989	−0.044
MAX	20.006	20.000	0.019	19.993	20.000	0.006	19.986	20.000	−0.001	19.973	20.000	−0.014	19.967	20.000	−0.020
MIN	19.985	19.987	−0.015	19.972	19.987	−0.028	19.965	19.987	−0.035	19.952	19.987	−0.048	19.946	19.987	−0.054
MAX	25.006	25.000	0.019	24.993	25.000	0.006	24.986	25.000	−0.001	24.973	25.000	−0.014	24.960	25.000	−0.027
MIN	24.985	24.987	−0.015	24.972	24.987	−0.028	24.965	24.987	−0.035	24.952	24.987	−0.048	24.939	24.987	−0.061
MAX	30.006	30.000	0.019	29.993	30.000	0.006	29.986	30.000	−0.001	29.973	30.000	−0.014	29.960	30.000	−0.027
MIN	29.985	29.987	−0.015	29.972	29.987	−0.028	29.965	29.987	−0.035	29.952	29.987	−0.048	29.939	29.987	−0.061
MAX	40.007	40.000	0.023	39.992	40.000	0.008	39.983	40.000	−0.001	39.966	40.000	−0.018	39.949	40.000	−0.035
MIN	39.982	39.984	−0.018	39.967	39.984	−0.033	39.958	39.984	−0.042	39.941	39.984	−0.059	39.924	39.984	−0.076
MAX	50.007	50.000	0.023	49.992	50.000	0.008	49.983	50.000	−0.001	49.966	50.000	−0.018	49.939	50.000	−0.045
MIN	49.982	49.984	−0.018	49.967	49.984	−0.033	49.958	49.984	−0.042	49.941	49.984	−0.059	49.914	49.984	−0.086
MAX	60.009	60.000	0.028	59.991	60.000	0.010	59.979	60.000	−0.002	59.958	60.000	−0.023	59.924	60.000	−0.057
MIN	59.979	59.981	−0.021	59.961	59.981	−0.039	59.949	59.981	−0.051	59.928	59.981	−0.072	55.894	59.981	−0.106
MAX	80.009	80.000	0.028	79.991	80.000	0.010	79.979	80.000	−0.002	79.952	80.000	−0.029	79.909	80.000	−0.072
MIN	79.979	79.981	−0.021	79.961	79.981	−0.039	79.949	79.981	−0.051	79.922	79.981	−0.078	79.879	79.981	−0.121
MAX	100.010	100.000	0.032	99.990	100.000	0.012	99.976	100.000	−0.002	99.942	100.000	−0.036	99.889	100.000	−0.089
MIN	99.975	99.978	−0.025	99.955	99.978	−0.045	99.941	99.978	−0.059	99.907	99.978	−0.093	99.854	99.978	−0.146
MAX	120.010	120.000	0.032	119.990	120.000	0.012	119.976	120.000	−0.002	119.934	120.000	−0.044	119.869	120.000	−0.109
MIN	119.975	119.978	−0.025	119.955	119.978	−0.045	119.941	119.978	−0.059	119.899	119.978	−0.101	119.834	119.978	−0.166
MAX	160.012	160.000	0.037	159.988	160.000	0.013	159.972	160.000	−0.003	159.915	160.000	−0.060	159.825	160.000	−0.150
MIN	159.972	159.975	−0.028	159.948	159.975	−0.052	159.932	159.975	−0.068	159.875	159.975	−0.125	159.785	159.975	−0.215
MAX	200.013	200.000	0.042	199.986	200.000	0.015	199.967	200.000	−0.004	199.895	200.000	−0.076	199.781	200.000	−0.190
MIN	199.967	199.971	−0.033	199.940	199.971	−0.060	199.921	199.971	−0.079	199.849	199.971	−0.151	199.735	199.971	−0.265
MAX	250.013	250.000	0.042	249.986	250.000	0.015	249.967	250.000	−0.004	249.877	250.000	−0.094	249.733	250.000	−0.238
MIN	249.967	249.971	−0.033	249.940	249.971	−0.060	249.921	249.971	−0.079	249.831	249.971	−0.169	149.687	249.971	−0.313
MAX	300.016	300.000	0.048	299.986	300.000	0.018	299.964	300.000	−0.004	299.850	300.000	−0.118	299.670	300.000	−0.298
MIN	299.964	299.968	−0.036	299.934	299.968	−0.066	299.912	299.968	−0.088	299.798	299.968	−0.202	299.618	299.968	−0.382
MAX	400.017	400.000	0.053	399.984	400.000	0.020	399.959	400.000	−0.005	399.813	400.000	−0.151	399.586	400.000	−0.378
MIN	399.960	399.964	−0.040	399.927	399.964	−0.073	399.902	399.964	−0.098	399.756	399.964	−0.244	399.529	399.964	−0.471
MAX	500.018	500.000	0.058	499.983	500.000	0.023	499.955	500.000	−0.005	499.771	500.000	−0.189	499.483	500.000	−0.477
MIN	499.955	499.960	−0.045	499.920	499.960	−0.080	499.892	499.960	−0.108	499.708	499.960	−0.292	499.420	499.960	−0.580

6.13 LIMIT DIMENSIONS—TOLERANCE ACCUMULATION

Whenever there is a chain or string of dimensions, the position of the dimensioned elements may vary as a result of the overall accumulation of individual tolerances. Chain dimensions produce the greatest tolerance accumulation. See Figure 6-22(a). Dimensioning from a common datum reduces tolerance accumulation since the tolerance on distance between any two elements involves only the two dimensions from the datum. See Figure 6-22(b). If the distance between two elements must be more closely controlled, the distance should be dimensioned directly to avoid tolerance accumulation. See Figure 6-22(c).

Tolerances may also be specified to control geometric form and true position of a feature in relation to a prescribed datum surface. Discussion of these methods is beyond the scope of this text, but may be found in ANSI Y14.5, Dimensioning and Tolerance for Engineering Drawings, published by the American Society of Mechanical Engineers.

Figure 6-22 Tolerance Accumulation

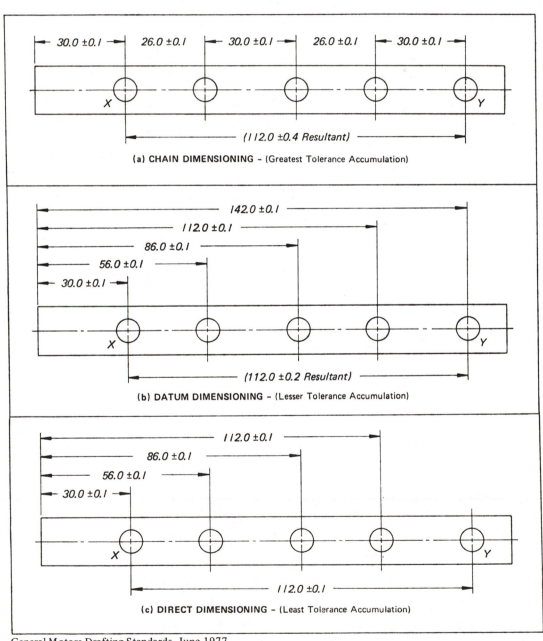

General Motors Drafting Standards, June 1977.

6.14 THREADED FASTENERS

Threaded fasteners are encountered in the design, fabrication, and operation of nearly all engineering products. Consequently, it is necessary to be able to recognize, represent, and specify screw threads and threaded fasteners on engineering drawings.

A *screw thread* is a helical groove around a cylinder. Figure 6-23 shows a typical vee-shaped screw thread. Square threads or trapezoidal threads are used to transmit power. Threads are *left-* or *right-hand*, depending on the direction of the helix. Right-hand screws advance when turned clockwise. Two, three, or more threads (multiple threads) can be cut side by side on a cylinder.

Figure 6-23 Screw Thread

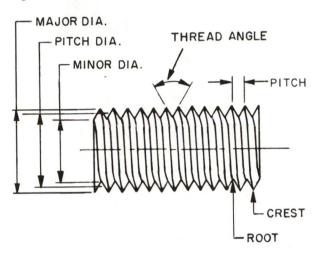

The true orthographic projections of threads are seldom drawn because they would project as helical curves and be difficult to construct. Therefore, symbols are used to facilitate the representation of threads and are of three types: detailed, schematic, and simplified. Figure 6-24 shows the end and side views of each type of thread representation. Note that the end view of the hole is the same for all types. If the external thread is chamfered, as in the simplified symbol, two visible circles are shown in the end view. Figure 6-26 shows the detailed symbol for a "Unified or American National" thread. This symbol will change if the thread form is changed (see the definition of *thread form*, Section 6.15). The schematic and simplified thread symbols are used for all thread forms.

6.15 THREAD DEFINITIONS

1. *External thread:* a thread on the outside of a cylinder.

2. *Internal thread:* a thread on the inside of a cylindrical hole.

3. *Major diameter:* the maximum diameter of an external or internal thread.

4. *Minor diameter:* the minimum diameter of an external or internal thread.

5. *Pitch diameter:* the diameter of an imaginary cylinder where the thread width is equal to the space between the threads.

6. *Pitch:* the distance from a point on one groove to the corresponding point on an adjacent groove.

7. *Crest:* the top or peak of a thread.

8. *Thread angle:* the angle between thread surfaces.

9. *Root:* the bottom of a thread.

10. *Thread form:* the shape of a thread. Common thread forms are: Sharp V (∧∧∧), Unified (ᒋᒋᒋ), Square (ⸯⵎⵎ), Acme (ⴖⴖⴖ), Buttress (ⴖⴖⴖ), and Knuckle (ⴖⴖⴖ).

11. *Thread series:* in the unified screw thread series, the classification of a thread which indicates the number of threads per inch for a particular diameter: for example, coarse, fine, extra fine, 8, 12, and 16 series. See Table 6-11, "Unified and American Screw Threads," at the end of the chapter. In the metric system, the classification of a thread is coarse or fine.

12. *Class of fit:* specifies the tightness of fit between external and internal threads in the unified screw thread series. Fit may be class 1, 2, or 3 (tightest) followed by an A (external) or a B (internal). In the metric system, tolerance classes 6g (external) and 6H (internal) are commonly used for general-purpose applications. These fits correspond to 2A and 2B of the unified series.

13. *Right-hand thread:* a thread that advances when rotated clockwise.

14. *Left-hand thread:* a thread that advances when rotated counterclockwise.

15. *Single thread:* one helical groove.

16. *Multiple thread:* two or more adjacent threads.

17. *Lead:* the distance that a thread will advance when rotated 360°. The lead for a single thread is equal to the pitch. For multiple threads the lead is equal to the pitch multiplied by the number of threads.

18. *Tapped hole:* a hole threaded with a special cutting tool called a tap.

Figure 6-24 Thread Representations

DETAILED REPRESENTATION

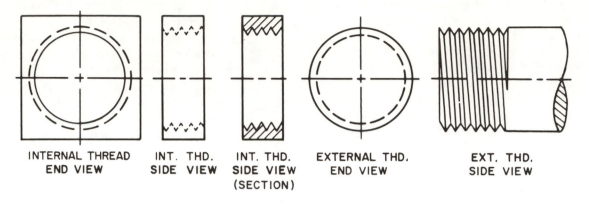

INTERNAL THREAD
END VIEW

INT. THD.
SIDE VIEW

INT. THD.
SIDE VIEW
(SECTION)

EXTERNAL THD.
END VIEW

EXT. THD.
SIDE VIEW

SCHEMATIC REPRESENTATION

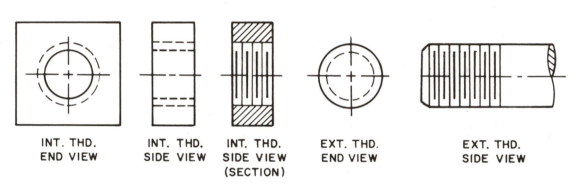

INT. THD.
END VIEW

INT. THD.
SIDE VIEW

INT. THD.
SIDE VIEW
(SECTION)

EXT. THD.
END VIEW

EXT. THD.
SIDE VIEW

SIMPLIFIED REPRESENTATION

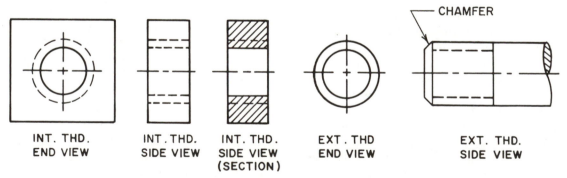

INT. THD.
END VIEW

INT. THD.
SIDE VIEW

INT. THD.
SIDE VIEW
(SECTION)

EXT. THD
END VIEW

CHAMFER

EXT. THD.
SIDE VIEW

6.16 THREAD SPECIFICATIONS

The most important part of a thread representation is the thread note, which gives the exact specifications of the thread. The thread symbol is used only for identification on the drawing.

Metric threads are specified by the letter "M" followed by the nominal size (basic major diameter) in millimeters and the pitch in millimeters separated by the symbol "X." The specifications may also be given by dimensioning the thread diameter (Figure 6–25).

Figure 6-25   Metric Thread Specifications

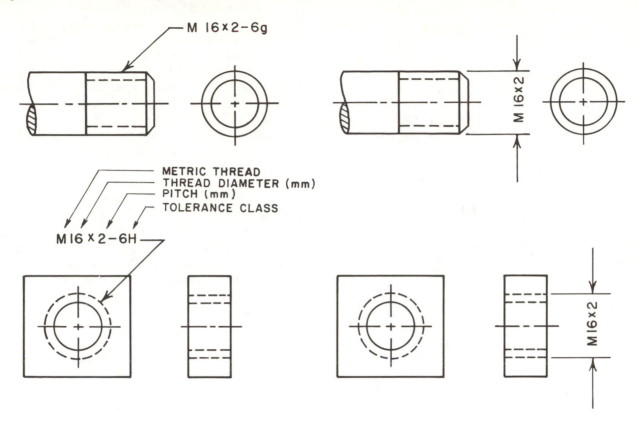

M 16×2-6g

METRIC THREAD
THREAD DIAMETER (mm)
PITCH (mm)
TOLERANCE CLASS

M 16 × 2 – 6H

M 16×2

M 16×2

Unified threads are specified by thread notes, as illustrated in Figure 6-26. The thread is understood to be single and right-hand unless otherwise noted.

Figure 6-26   Unified Thread Specifications

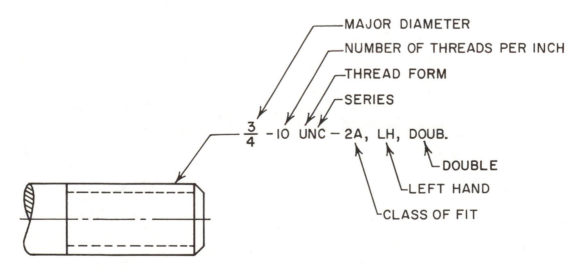

MAJOR DIAMETER
NUMBER OF THREADS PER INCH
THREAD FORM
SERIES

$\frac{3}{4}$ – 10 UNC – 2A, LH, DOUB.

DOUBLE
LEFT HAND
CLASS OF FIT

A complete fastener note includes the thread note followed by the length, material, type of head, and designation of the particular type of fastener (Figure 6-27).

Figure 6-27 Complete Fastener Note

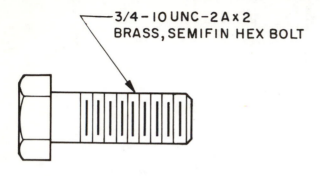

Tapped holes are specified in the unified system by giving, in note form, the tap drill size and depth followed by the thread specification and depth of thread. In the metric system, dimensions are generally provided. Figure 6-28 shows typical tapped hole specifications in both the unified and metric systems.

The depth of tapped holes and the entrance length of threaded fasteners vary depending on size, type of thread, and material. A good proportion for general use is a hole depth equal to twice the thread diameter and a thread depth equal to three-fourths of the hole depth.

6.17 DRAWING A DETAILED THREAD REPRESENTATION

Detailed thread representations are normally used when the major thread diameter on the drawing is greater than 25 mm (1 in.). Figure 6-29 gives the step-by-step construction of a $1\frac{1}{8}$ -7UNC-2A thread.

The first step is to box in the area of the thread with construction lines. Next, beginning at the thread end, graphically divide 1 in. along the outline of the box into seven equal spaces (the pitch) and continue to set these distances off until the thread-length line is reached. Then set off one-half the pitch (for a single thread) along the opposite side of the thread box. Connect the upper left corner to this point with a straight line and continue to draw crest lines parallel to this line through the pitch points by sliding the triangle until the thread-length line is reached. Construct 60° V's through the points where the crest lines touch the box outline. Construct root lines between the inward vertices of the V's. Terminate the thread near the thread-length line, and the symbol is complete. Similar construction is used for other thread forms.

6.18 SCHEMATIC AND SIMPLIFIED THREAD REPRESENTATIONS

Schematic and simplified thread symbols are constructed by using pitch and thread depth to maintain the proportions of the symbols illustrated in Figure 6-24.

It is recommended that these symbols be used for showing threads under 25 mm in diameter, drawing size. Also, the schematic symbol is generally used on assembly drawings, and the simplified symbol is most often used on detail drawings.

Figure 6-28 Tapped Hole Specifications

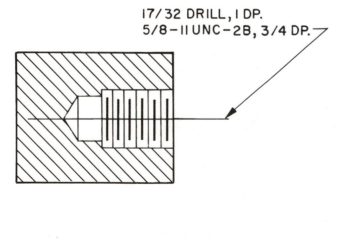

UNIFIED

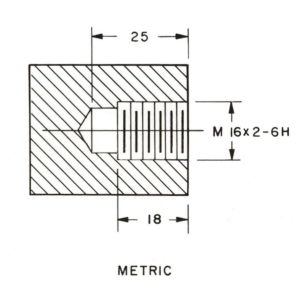

METRIC

Figure 6-29 Drawing a Detailed Thread Representation

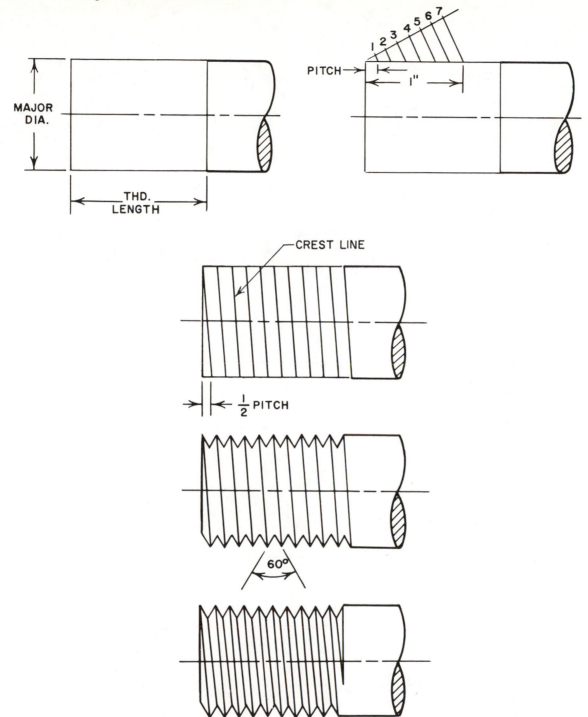

The two sets of symbols should be compared and the similarities and differences noted. The schematic and simplified symbols are the same for hidden threads. The end-view symbols for external threads are the same, but the end view of an external thread differs from the end view of an internal thread.

It is not necessary to show the actual pitch of the thread or the depth of thread by spacing the lines of the symbol. Symbols for several threads of the same diameter and different pitch may be shown by the same symbol. The symbols should be made to look good and to be easy to read, without regard for other considerations.

107

6.19 THREADED FASTENERS—GENERAL

Many types of threaded fasteners have been developed for specific purposes. These include bolts and nuts of various descriptions, studs, cap screws, machine screws, set screws, and wood screws. Additionally, these fasteners may have different types or shapes of heads, such as hexagonal, round, fillister, socket, etc., which are illustrated in manufacturers' catalogues. Complete specifications for fasteners can be found in tables such as "Metric Hex Cap Screws and Hex Bolts" (Table 6-9) and "Dimensions of Hex Bolts" (Table 6-12).

6.20 FASTENERS—GENERAL

Many fasteners besides the threaded fasteners discussed in this chapter are used in engineering design and fabrication. Some of these which can be classified as removable fasteners are taper pins, snap rings, keys, and springs. Others that may be classified as permanent fasteners are rivets, welds, and adhesives. Complete information on all types of fasteners may be found in standard reference tables similar to Tables 6-7 through 6-13.

Table 6-7 Metric Thread Series

Only the most commonly used metric thread series are given here.

Basic Major Diameter			Coarse	Fine Thread		
Choice 1 (mm)	Choice 2 (mm)	Choice 3 (mm)	Pitch (mm)	Pitch (mm)		
1.6	-	-	0.35		0.2	
-	1.8	-	0.35		0.2	
2.0	-	-	0.4		0.25	
-	2.2	-	0.45		0.25	
2.5	-	-	0.45		0.35	
3	-	-	0.5		0.35	
-	3.5	-	0.6		0.35	
4	-	-	0.7		0.5	
-	4.5	-	0.75		0.5	
5	-	-	0.8		0.5	
-	-	5.5	-		0.5	
6	-	-	1		0.75	
-	-	7	1		0.75	
8	-	-	1.25		1	0.75
-	-	9	1.25		1	0.75
10	-	-	1.5	1.25	1	0.75
-	-	11	1.5		1	0.75
12	-	-	1.75	1.5	1.25	1
-	14	-	2	1.5	1.25	1
-	-	15	-		1.5	1
16	-	-	2		1.5	1
-	-	17	-		1.5	1
-	18	-	2.5	2	1.5	1
20	-	-	2.5	2	1.5	1
-	22	-	2.5	2	1.5	1
24	-	-	3	2	1.5	1

Table 6–8 Recommended Tap Drill Diameters for Metric Threads

The recommendations for drilling prior to conventional tapping to produce internal metric screw threads are given below. The recommended drill sizes are in agreement with ISO 2306, Drills for Use Prior to Tapping Screw Threads.

Nominal Size	Internal Thread Minor Dia		Tap Drill Dia	Nominal Size	Internal Thread Minor Dia		Tap Drill Dia
	Max	Min			Max	Min	
M1.6x0.35	1.321	1.221	1.25	M16x2	14.210	13.835	14.0
M2x0.4	1.679	1.567	1.6	M20x2.5	17.744	17.294	17.5
M2.5x0.45	2.138	2.013	2.05	M24x3	21.252	20.752	21.0
M3x0.5	2.599	2.459	2.5	M30x3.5	27.771	26.211	26.5
M3.5x0.6	3.010	2.850	2.9	M36x4	32.270	31.670	32.0
M4x0.7	3.422	3.242	3.3	M42x4.5	37.799	37.129	37.5
M5x0.8	4.334	4.134	4.2	M48x5	43.297	42.587	43.0
M6.3x1	5.553	5.217	5.3	M56x5.5	50.796	50.046	50.5
M8x1.25	6.912	6.647	6.8	M64x6	58.305	57.505	58.0
M10x1.5	8.676	8.376	8.5	M72x6	66.305	65.505	66.0
M12x1.75	10.441	10.106	10.2	M80x6	74.305	73.505	74.0
M14x2	12.210	11.835	12.0	M90x6	84.305	83.505	84.0
				M100x6	94.305	93.505	94.0

NOTE: All dimensions are millimeters.

Table 6-9 Metric Hex Cap Screws and Hex Bolts

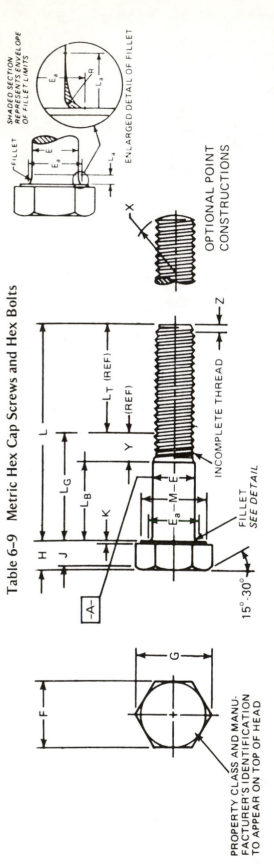

PROPERTY CLASS AND MANUFACTURER'S IDENTIFICATION TO APPEAR ON TOP OF HEAD

OPTIONAL POINT CONSTRUCTIONS

ENLARGED DETAIL OF FILLET

SHADED SECTION REPRESENTS ENVELOPE OF FILLET LIMITS

Nom Screw Size & Thread Pitch	E Body Diameter Max	E Body Diameter Min	F Width Across Flats Max	F Width Across Flats Min	G Width Across Corners Max	G Width Across Corners Min	H Head Height Max	H Head Height Min	J Wrenching Height Min	K Washer Face Thickness Max	K Washer Face Thickness Min	M Washer Face Dia Min	Runout of Bearing Surface FIR Max	Eₐ Fillet Transition Dia Max	Lₐ Fillet Transition Length Max	R Radius of Fillet Min	Lₜ (Ref) Thread Length, basic Screw Lengths ≤125	Lₜ (Ref) Thread Length, basic Screw Lengths >125 and ≤200	Lₜ (Ref) Thread Length, basic Screw Lengths >200	Y (Ref) Transition Thread Length Max
M5x0.8	5.00	4.82	8.00	7.78	9.24	8.87	3.65	3.35	2.4	0.5	0.2	7.0	0.22	5.7	1.2	0.2	16	22	35	4.0
M6.3x1	6.30	6.08	10.00	9.76	11.55	11.13	4.47	4.13	3.0	0.5	0.2	8.9	0.25	7.3	1.8	0.3	18.6	24.6	37.6	5.0
M8x1.25	8.00	7.78	13.00	12.73	15.01	14.51	5.50	5.10	3.7	0.6	0.3	11.6	0.28	9.2	2.0	0.4	22	28	41	6.2
M10x1.5	10.00	9.78	15.00	14.70	17.32	16.76	6.63	6.17	4.5	0.6	0.3	13.6	0.31	11.2	2.0	0.4	26	32	45	7.5
M12x1.75	12.00	11.73	18.00	17.67	20.78	20.14	7.76	7.24	5.2	0.6	0.3	16.6	0.35	13.2	3.0	0.4	30	36	49	8.8
M14x2	14.00	13.73	21.00	20.64	24.25	23.53	9.09	8.51	6.2	0.6	0.3	19.4	0.39	15.2	3.0	0.6	34	40	53	10.0
M16x2	16.00	15.73	24.00	23.61	27.71	26.92	10.32	9.68	7.0	0.8	0.4	22.4	0.43	17.7	3.0	0.6	38	44	57	10.0
M20x2.5	20.00	19.67	30.00	29.35	34.64	33.46	12.88	12.12	8.8	0.8	0.4	27.6	0.53	22.4	4.0	0.8	46	52	65	12.5
M24x3	24.00	23.67	36.00	35.25	41.57	40.19	15.44	14.56	10.5	0.8	0.4	32.9	0.63	26.4	4.0	0.8	54	60	73	15.0
M30x3.5	30.00	29.61	46.00	44.50	53.12	50.73	19.48	17.92	13.1	0.8	0.4	42.5	0.78	33.4	6.0	1.0	66	72	85	17.5
M36x4	36.00	35.61	55.00	53.20	63.51	60.65	23.38	21.62	15.8	0.8	0.4	50.8	0.93	39.4	6.0	1.0	78	84	97	20.0
M42x4.5	42.00	41.61	65.00	62.90	75.06	71.71	26.97	25.03	18.2	1.0	0.5	58.5	1.09	45.6	6.3	1.2	90	96	109	22.5
M48x5	48.00	47.61	75.00	72.60	86.60	82.76	31.07	28.93	21.0	1.0	0.5	67.5	1.25	52.6	8.0	1.5	102	108	121	25.0
M56x5.5	56.00	55.54	85.00	82.20	98.15	93.71	36.20	33.80	24.5	1.0	0.5	76.5	1.47	62.0	10.5	2.0	—	124	137	27.5
M64x6	64.00	63.54	95.00	91.80	109.70	104.65	41.32	38.68	28.0	1.0	0.5	85.5	1.69	70.0	10.5	2.0	—	140	153	30.0
M72x6	72.00	71.54	105.00	101.40	121.24	115.60	46.45	43.55	31.5	1.2	0.6	94.5	1.91	78.0	10.5	2.0	—	156	169	30.0
M80x6	80.00	79.54	115.00	111.00	132.79	126.54	51.58	48.42	35.0	1.2	0.6	103.5	2.13	86.0	10.5	2.0	—	172	185	30.0
M90x6	90.00	89.46	130.00	125.50	150.11	143.07	57.74	54.26	39.2	1.2	0.6	117.0	2.41	96.0	10.5	2.0	—	192	205	30.0
M100x6	100.00	99.46	145.00	140.00	167.43	159.60	63.90	60.10	43.4	1.2	0.6	130.5	2.69	107.0	12.2	2.5	—	212	225	30.0

International Fasteners Institute, 506, 1976.

110

Table 6-10 Metric Hex Nuts

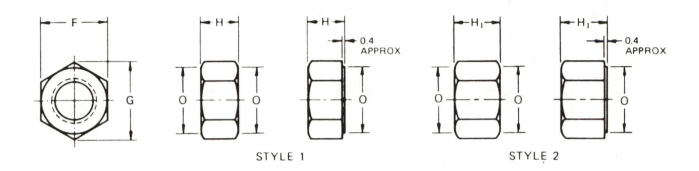

STYLE 1   STYLE 2

Nominal Nut Size and Thread Pitch	F Width Across Flats		G Width Across Corners		O Bearing Face Dia	H Nut Thickness Style 1		H₁ Nut Thickness Style 2		Runout of Bearing Surface FIR
	Max	Min	Max	Min	Min	Max	Min	Max	Min	Max
M1.6x0.35	3.20	3.02	3.70	3.44	2.5	—	—	1.3	1.1	—
M2x0.4	4.00	3.82	4.62	4.35	3.1	—	—	1.6	1.3	—
M2.5x0.45	5.00	4.82	5.77	5.49	4.1	—	—	2.0	1.7	—
M3x0.5	5.50	5.32	6.35	6.06	4.6	—	—	2.4	2.1	—
M3.5x0.6	7.00	6.78	8.08	7.73	6.0	—	—	2.8	2.5	—
M4x0.7	7.00	6.78	8.08	7.73	6.0	—	—	3.2	2.9	—
M5x0.8	8.00	7.78	9.24	8.87	7.0	4.5	4.2	5.3	5.0	0.30
M6.3x1	10.00	9.76	11.55	11.13	8.9	5.6	5.3	6.5	6.2	0.33
M8x1.25	13.00	12.73	15.01	14.51	11.6	6.6	6.2	7.8	7.4	0.36
M10x1.5	15.00	14.70	17.32	16.76	13.6	9.0	8.5	10.7	10.2	0.39
M12x1.75	18.00	17.67	20.78	20.14	16.6	10.7	10.2	12.8	12.3	0.42
M14x2	21.00	20.64	24.25	23.53	19.4	12.5	11.9	14.9	14.3	0.45
M16x2	24.00	23.61	27.71	26.92	22.4	14.5	13.9	17.4	16.8	0.48
M20x2.5	30.00	29.00	34.64	33.06	27.6	18.4	17.4	21.2	20.2	0.56
M24x3	36.00	34.80	41.57	39.67	32.9	22.0	20.9	25.4	24.3	0.64
M30x3.5	46.00	44.50	53.12	50.73	42.5	26.7	25.4	31.0	29.7	0.76
M36x4	55.00	53.20	63.51	60.65	50.8	32.0	30.5	37.6	36.1	0.89

International Fasterners Institute, 507, 1976.

111

Table 6–11 Unified and American Screw Threads

Basic Dimensions for Coarse, Fine, Extra-Fine, 8, 12, and 16 Thread Series.

Nominal Diam.	Basic Major Diam.	Coarse UNC, NC Classes 1A, 1B, 2A, 2B, 3A, 3B, 2, 3		Fine UNF, NF Classes 1A, 1B, 2A, 2B, 3A, 3B, 2, 3		Extra-Fine UNEF, NEF Classes 2A, 2B 2, 3		8 Thd. 8N Classes 2A, 2B 2, 3		12 Thd. 12UN, 12N Classes 2A, 2B 2, 3		16 Thd. 16UN, 16N Classes 2A, 2B 2, 3	
		Thds. per in.	Tap Drill	Thds. per in.	Tap Drill	Thds. per in.	Tap Drill	Thds. per in.	Tap Drill	Thds. per in.	Tap Drill	Thds. per in.	Tap Drill
# 0	0.0600	--	--	80	3/64								
# 1	0.0730	64	53	72	53								
# 2	0.0860	56	50	64	50								
# 3	0.0990	48	47	56	45								
# 4	0.1120	40	43	48	42								
# 5	0.1250	40	38	44	37								
# 6	0.1380	32	36	40	33								
# 8	0.1640	32	29	36	29								
#10	0.1900	24	25	32	21								
#12	0.2160	24	16	28	14	32	13						
1/4	0.2500	20	7	28	3	32	7/32						
5/16	0.3125	18	F	24	I	32	9/32						
3/8	0.3750	16	5/16	24	Q	32	11/32						
7/16	0.4375	14	U	20	25/64	28	13/32						
1/2	0.5000	13	27/64	20	29/64	28	15/32	--	--	12	27/64		
9/16	0.5625	12	31/64	18	33/64	24	33/64	--	--	12	31/64		
5/8	0.6250	11	17/32	18	37/64	24	37/64	--	--	12	35/64		
11/16	0.6875	--	--	--	--	24	41/64	--	--	12	39/64		
3/4	0.7500	10	21/32	16	11/16	20	45/64	--	--	12	43/64	16	11/16
13/16	0.8125	--	--	--	--	20	49/64	--	--	12	47/64	16	3/4
7/8	0.8750	9	49/64	14	13/16	20	53/64	--	--	12	51/64	16	13/16
15/16	0.9375	--	--	--	--	20	57/64	--	--	12	55/64	16	7/8
1	1.0000	--	--	14	15/16	--	--	8	7/8				
1	1.0000	8	7/8	12	59/64	20	61/64	--	--	12	59/64	16	15/16
1 1/16	1.0625	--	--	--	--	18	1	--	--	12	63/64	16	1
1 1/8	1.1250	7	63/64	12	1 3/64	18	1 5/64	8	1	12	1 3/64	16	1 1/16
1 3/16	1.1875	--	--	--	--	18	1 9/64	--	--	12	1 7/64	16	1 1/8
1 1/4	1.2500	7	1 7/64	12	1 11/64	18	1 3/16	8	1 1/8	12	1 11/64	16	1 3/16
1 5/16	1.3125	--	--	--	--	18	1 17/64	--	--	12	1 15/64	16	1 1/4
1 3/8	1.3750	6	1 7/32	12	1 19/64	18	1 5/16	8	1 1/4	12	1 19/64	16	1 5/16
1 7/16	1.4375	--	--	--	--	18	1 3/8	--	--	12	1 23/64	16	1 3/8
1 1/2	1.5000	6	1 11/32	12	1 27/64	18	1 7/16	8	1 3/8	12	1 27/64	16	1 7/16
1 9/16	1.5625	--	--	--	--	18	1 1/2	--	--	--	--	16	1 1/2
1 5/8	1.6250	--	--	--	--	18	1 9/16	8	1 1/2	12	1 35/64	16	1 9/16
1 11/16	1.6875	--	--	--	--	18	1 5/8	--	--	--	--	16	1 5/8
1 3/4	1.7500	5	1 9/16	--	--	16	1 11/16	8	1 5/8	12	1 43/64	16	1 11/16
1 13/16	1.8125	--	--	--	--	--	--	--	--	--	--	16	1 3/4
1 7/8	1.8750	--	--	--	--	--	--	8	1 3/4	12	1 51/64	16	1 13/16
1 15/16	1.9375	--	--	--	--	--	--	--	--	--	--	16	1 7/8
2	2.0000	4 1/2	1 25/32	--	--	16	1 15/16	8	1 7/8	12	1 59/64	16	1 15/16
2 1/16	2.0625	--	--	--	--	--	--	--	--	--	--	16	2
2 1/8	2.1250	--	--	--	--	--	--	8	2	12	2 3/64	16	2 1/16
2 3/16	2.1875	--	--	--	--	--	--	--	--	--	--	16	2 1/8
2 1/4	2.2500	4 1/2	2 1/32	--	--	--	--	8	2 1/8	12	2 11/64	16	2 3/16
2 5/16	2.3125	--	--	--	--	--	--	--	--	--	--	16	2 1/4
2 3/8	2.3750	--	--	--	--	--	--	--	--	12	2 19/64	16	2 5/16
2 7/16	2.4375	--	--	--	--	--	--	--	--	--	--	16	2 3/8
2 1/2	2.5000	4	2 1/4	--	--	--	--	8	2 3/8	12	2 27/64	16	2 7/16
2 5/8	2.6250	--	--	--	--	--	--	--	--	12	2 35/64	16	2 9/16
2 3/4	2.7500	4	2 1/2	--	--	--	--	8	2 5/8	12	2 43/64	16	2 11/16
2 7/8	2.8750	--	--	--	--	--	--	--	--	12	2 51/64	16	2 13/16
3	3.0000	4	2 3/4	--	--	--	--	8	2 7/8	12	2 59/64	16	2 15/16
3 1/8	3.1250	--	--	--	--	--	--	--	--	12	3 3/64	16	3 1/16
3 1/4	3.2500	4	3	--	--	--	--	8	3 1/8	12	3 11/64	16	3 3/16
3 3/8	3.3750	--	--	--	--	--	--	--	--	12	3 19/64	16	3 5/16
3 1/2	3.5000	4	3 1/4	--	--	--	--	8	3 3/8	12	3 27/64	16	3 7/16
3 5/8	3.6250	--	--	--	--	--	--	--	--	12	3 35/64	16	3 9/16
3 3/4	3.7500	4	3 1/2	--	--	--	--	8	3 5/8	12	3 43/64	16	3 11/16
3 7/8	3.8750	--	--	--	--	--	--	--	--	12	3 51/64	16	3 13/16
4	4.0000	4	3 3/4	--	--	--	--	8	3 7/8	12	3 59/64	16	3 15/16

Compiled from ANSI B1.1–1960.
BOLDFACE TYPE indicates products unified dimensionally with British and Canadian Standards.

Table 6-12 Dimensions of Hex Bolts

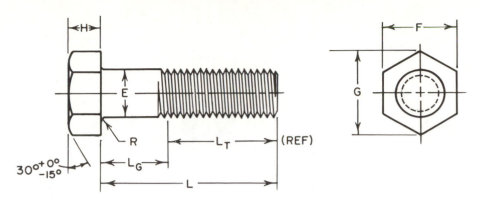

Nominal Size or Basic Product Dia.	E Body Dia. Max	F Width Across Flats			G Width Across Corners		H Height			R Radius of Fillet		L_T Thread Length for Bolt Lengths	
		Basic	Max	Min	Max	Min	Basic	Max	Min	Max	Min	6 in. and Shorter Basic	Over 6 in. Basic
1/4	0.2500	0.260	7/16	0.438	0.425	0.505	0.484	11/64	0.188	0.150	0.03	0.01	0.750	1.000
5/16	0.3125	0.324	1/2	0.500	0.484	0.577	0.552	7/32	0.235	0.195	0.03	0.01	0.875	1.125
3/8	0.3750	0.388	9/16	0.562	0.544	0.650	0.620	1/4	0.268	0.226	0.03	0.01	1.000	1.250
7/16	0.4375	0.452	5/8	0.625	0.603	0.722	0.687	19/64	0.316	0.272	0.03	0.01	1.125	1.375
1/2	0.5000	0.515	3/4	0.750	0.725	0.866	0.826	11/32	0.364	0.302	0.03	0.01	1.250	1.500
5/8	0.6250	0.642	15/16	0.938	0.906	1.083	1.033	27/64	0.444	0.378	0.06	0.02	1.500	1.750
3/4	0.7500	0.768	1-1/8	1.125	1.088	1.299	1.240	1/2	0.524	0.455	0.06	0.02	1.750	2.000
7/8	0.8750	0.895	1-5/16	1.312	1.269	1.516	1.447	37/63	0.604	0.531	0.06	0.02	2.000	2.250
1	1.0000	1.022	1-1/2	1.500	1.450	1.732	1.653	1-43/64	0.700	0.591	0.09	0.03	2.250	2.500
1-1/8	1.1250	1.149	1-11/16	1.688	1.631	1.949	1.859	3/4	0.780	0.658	0.09	0.03	2.500	2.750
1-1/4	1.2500	1.277	1-7/8	1.875	1.812	2.165	2.066	27/32	0.876	0.749	0.09	0.03	2.750	3.000
1-3/8	1.3750	1.404	2-1/16	2.062	1.994	2.382	2.273	29/32	0.940	0.810	0.09	0.03	3.000	3.250
1-1/2	1.5000	1.531	2-1/4	2.250	2.175	2.598	2.480	1	1.036	0.902	0.09	0.03	3.250	3.500
1-3/4	1.7500	1.785	2-5/8	2.625	2.538	3.031	2.893	1-5/32	1.196	1.054	0.12	0.04	3.750	4.000
2	2.0000	2.039	3	3.000	2.900	3.464	3.306	1-11/32	1.388	1.175	0.12	0.04	4.250	4.500
2-1/4	2.2500	2.305	3-3/8	3.375	3.262	3.897	3.719	1-1/2	1.548	1.327	0.19	0.06	4.750	5.000
2-1/2	2.5000	2.559	3-3/4	3.750	3.625	4.330	4.133	1-21/32	1.708	1.479	0.19	0.06	5.250	5.500
2-3/4	2.7500	2.827	4-1/8	4.125	3.988	4.763	4.546	1-13/16	1.869	1.632	0.19	0.06	5.750	6.000
3	3.0000	3.081	4-1/2	4.500	4.350	5.196	4.959	2	2.060	1.815	0.19	0.06	6.250	6.500
3-1/4	3.2500	3.335	4-7/8	4.875	4.712	5.629	5.372	2-3/16	2.251	1.936	0.19	0.06	6.750	7.000
3-1/2	3.5000	3.589	5-1/4	5.250	5.075	6.062	5.786	2-5/16	2.380	2.057	0.19	0.06	7.250	7.500
3-3/4	3.7500	3.858	5-5/8	5.625	5.437	6.495	6.198	2-1/2	2.572	2.241	0.19	0.06	7.750	8.000
4	4.0000	4.111	6	6.000	5.800	6.928	6.612	2-11/16	2.764	2.424	0.19	0.06	8.250	8.500

Courtesy of ANSI; B18.2.1–1972.

113

Table 6-13 Dimensions of Hex Nuts and Jam Nuts

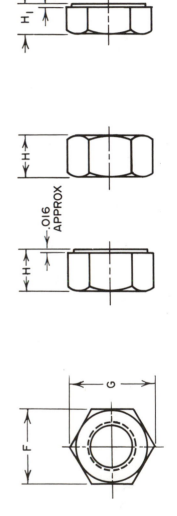

Nominal Size or Basic Major Dia. of Thread		F Width Across Flats			G Width Across Corners		H Thickness Hex Nuts			H₁ Thickness Hex Jam Nuts			Hex Nuts Specified Proof Load Runout of Bearing Face, FIR Max		Jam Nuts All Strength Nuts
		Basic	Max	Min	Max	Min	Basic	Max	Min	Basic	Max	Min	Up to 150,000 psi	150,000 psi and Greater	Runout of Bearing Face, FIR Max
1/4	0.2500	7/16	0.438	0.428	0.505	0.488	7/32	0.226	0.212	5/32	0.163	0.150	0.015	0.010	0.015
5/16	0.3125	1/2	0.500	0.489	0.577	0.557	17/64	0.273	0.258	3/16	0.195	0.180	0.016	0.011	0.016
3/8	0.3750	9/16	0.562	0.551	0.650	0.628	21/64	0.337	0.320	7/32	0.227	0.210	0.017	0.012	0.017
7/16	0.4375	11/16	0.688	0.675	0.794	0.768	3/8	0.385	0.365	1/4	0.260	0.240	0.018	0.013	0.018
1/2	0.5000	3/4	0.750	0.736	0.866	0.840	7/16	0.448	0.427	5/16	0.323	0.302	0.019	0.014	0.019
9/16	0.5625	7/8	0.875	0.861	1.010	0.982	31/64	0.496	0.473	5/16	0.324	0.301	0.020	0.015	0.020
5/8	0.6250	15/16	0.938	0.922	1.083	1.051	35/64	0.559	0.535	3/8	0.387	0.363	0.021	0.016	0.021
3/4	0.7500	1-1/8	1.125	1.088	1.299	1.240	41/64	0.665	0.617	27/64	0.446	0.398	0.023	0.018	0.023
7/8	0.8750	1-5/16	1.312	1.269	1.516	1.447	3/4	0.776	0.724	31/64	0.510	0.458	0.025	0.020	0.025
1	1.0000	1-1/2	1.500	1.450	1.732	1.653	55/64	0.887	0.831	35/64	0.575	0.519	0.027	0.022	0.027
1-1/8	1.1250	1-11/16	1.688	1.631	1.949	1.859	31/32	0.939	0.999	39/64	0.639	0.579	0.030	0.025	0.030
1-1/4	1.2500	1-7/8	1.875	1.812	2.165	2.066	1-1/16	1.094	1.030	23/32	0.751	0.687	0.033	0.028	0.033
1-3/8	1.3750	2-1/16	2.062	1.994	2.382	2.273	1-11/64	1.206	1.138	25/32	0.815	0.747	0.036	0.031	0.036
1-1/2	1.5000	2-1/4	2.250	2.175	2.598	2.480	1-9/32	1.317	1.245	27/32	0.880	0.808	0.039	0.034	0.039

Courtesy of ANSI; B18.2.2–1972.

114

Engineering Drawings

7

Engineering drawings are used to describe the design, manufacture, or assembly of engineering products or projects. They are found in many forms, depending upon the engineering discipline in which they are used. An electrical drawing may be in the form of a schematic or wiring diagram; a civil engineering drawing may be a topographic map or a structural steel drawing; a chemical engineering drawing may be a piping layout, while in mechanical engineering it may be the welding drawing of a fabricated machine part.

7.1 WORKING DRAWINGS

Working drawings are generally defined as drawings that convey information for the fabrication of machinery, products, or structures. Thousands of drawings of each part, assembly, system, and supporting equipment of a modern jet aircraft would constitute a complete set of working drawings. A set of working drawings consists of detail, subassembly, and assembly drawings. Each drawing must have a descriptive title, and assembly drawings should include a parts list or bill of materials.

Working drawings can be made on a wide variety of drawing paper, cloth, and plastic. Drawing sheet sizes have been standardized and come in incremental sizes

starting with the smallest size, A ($8\frac{1}{2} \times 11$ in.), to the largest size, E (34×44 in.). Size B is 11×17 in., size C is 17×22 in., and size D is 22×34 in. Drawing material is also available in rolls for very large drawings.

All drawings have a title block, usually preprinted on the sheet, which includes all the pertinent information for the drawings produced by any individual company. The title block shown in Figure 7-1 is typical of a preprinted title block. A typical materials list is illustrated in Figure 7-4.

Although some drawings are inked as original master copies, most engineering drawings are done in pencil. Sharp, dark pencil work will reproduce with excellent results using modern printing processes.

Figure 7-1 Title Block

Figure 7-2 First- and Third-Angle Projection

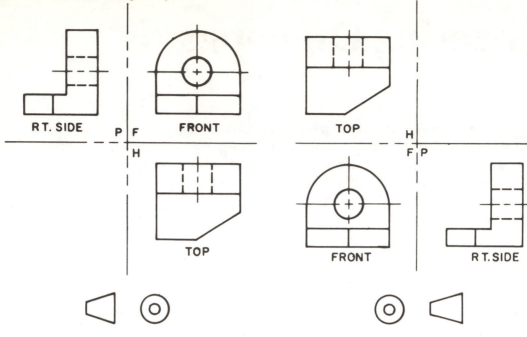

FIRST-ANGLE PROJECTION   THIRD-ANGLE PROJECTION

Figure 7-3 Detail Drawing

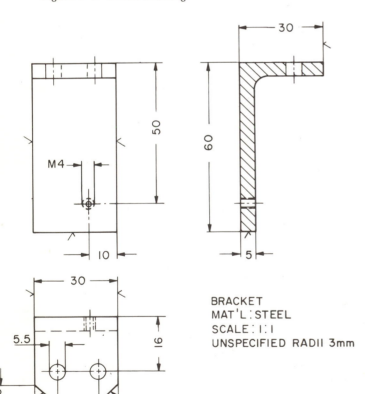

BRACKET
MAT'L : STEEL
SCALE : 1 : 1
UNSPECIFIED RADII 3mm

In the United States, the conventional multiview projections used for engineering drawings are referred to as *third-angle projections*. In Europe, the conventional system is *first-angle projection*. This difference between U.S. and European drawings is mentioned because European drawings are also used in this country. The two systems are illustrated in Figure 7-2. Note the use of the symbol ☐ ⊙ to denote first-angle projection and the symbol ⊙ ☐ to denote third-angle projection.

7.2 DETAIL DRAWINGS

A complete drawing of a single part containing the necessary information to manufacture that part is called a *detail drawing*. The detail drawing must contain a sufficient number of views to give a complete shape description of the part. The views must be dimensioned carefully to include all the necessary size and location dimensions needed for manufacturing the part. Finish marks, shop notes, material specifications, kind of finish, references to related drawings, and any other information deemed necessary to assure that the detail drawing is a complete manufacturing document must be included. It is the final duty of the engineer responsible for a project to add his signature to the detail drawing below the signature of the draftsman and checker to certify that he assumes full responsibility. A detail drawing complete with subtitle is shown in Figure 7-3. This drawing is in first-angle projection showing front, top, and left side views.

Figure 7-4 Subassembly Drawing

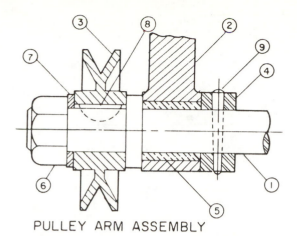

PULLEY ARM ASSEMBLY

9	PIN	I	STEEL	# 3 X $\frac{3}{4}$ T'PR
8	KEY	I	STEEL	# 506 W.F.
7	WASHER	I	STEEL	$\frac{13}{16}$ I.D. H'V'Y.
6	NUT	I	STEEL	REG $\frac{3}{4}$ - 12
5	BUSHING	I	BRONZE	
4	COLLAR	I	STEEL	
3	SHEAVE	I	CAST I.	
2	ARM	I	CAST I.	
I	SHAFT	I	STEEL	
NO.	NAME	REQ	MATERIAL	NOTES

Figure 7-5 Detail Assembly Drawing

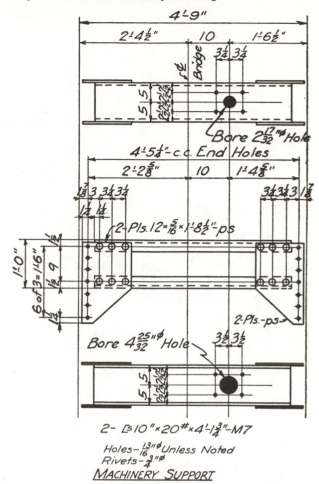

7.3 ASSEMBLY DRAWINGS

Drawings that show the relative positions of the various parts of a machine or structure are called *assembly drawings.* Types of assembly drawings include *unit* or *subassembly drawings,* which illustrate one unit of a complicated machine, such as the carburetor for an automobile engine. *Exploded assembly drawings* show the pieces of a unit separated in proper sequence for assembly. *Detail assembly drawings,* as the name implies, give all the information necessary to construct and assemble the machine or structure. See Figures 7-4, 7-5, and 7-6. Other types include design assembly, working assembly, layout, and installation drawings and diagrams.

Figure 7-6 Exploded Assembly Drawing

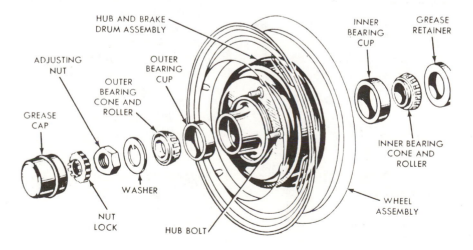

117

Figure 7-7 Piping Drawings

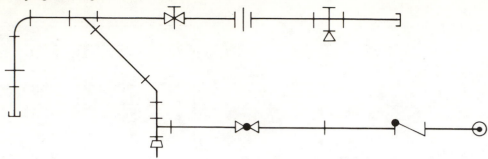

SINGLE LINE DRAWING

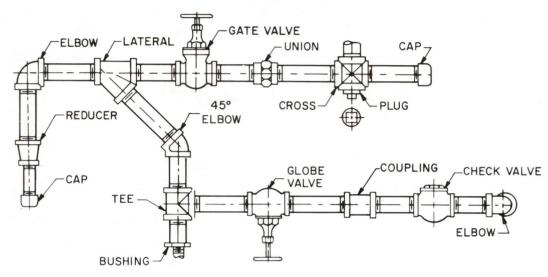

DOUBLE-LINE DRAWING

Figure 7-8 Welding Drawing

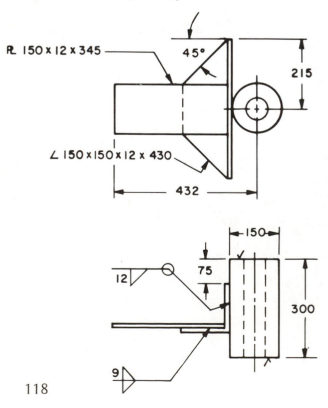

7.4 DISTINCTIVE TYPES OF ENGINEERING DRAWINGS

Although most engineering drawings can be classified as working drawings, many are more easily identified or classified by their unique use, such as architectural drawings, electrical drawings, and topographic drawings. The more common of these distinctive types of drawings are illustrated on the following pages. American National Standards Institute (ANSI) symbols may be found in most engineering reference manuals.

Piping drawings (Figure 7-7) can be single-line drawings that show only symbols for the pipes, fittings, and joints. They can also be double-line drawings or true orthographic projections.

Welding drawings (Figure 7-8) are similar to other machine drawings, except that unique welding symbols are used.

Electrical drawings are most often found in the form of schematic or symbolized diagrams. However, circuit-board layouts, parts layouts, and electrical component drawings are frequently used in the design of electronic systems. Figure 7-9 illustrates different types of electronic drawings.

118

Figure 7-9 Electronic Drawings

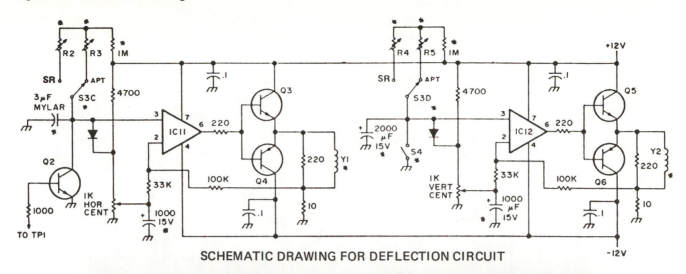

SCHEMATIC DRAWING FOR DEFLECTION CIRCUIT

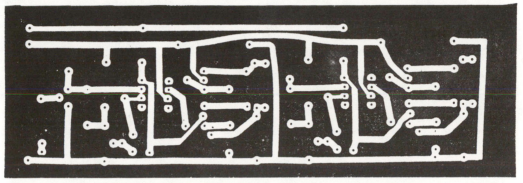

PRINTED CIRCUIT BOARD FOR DEFLECTION CIRCUIT

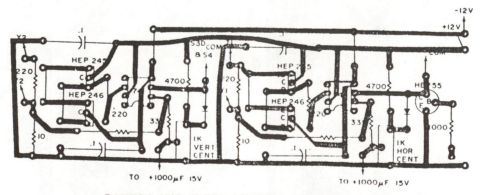

PARTS LAYOUT FOR PRINTED CIRCUIT BOARD

DEFLECTION CIRCUIT ASSEMBLED

Courtesy *73 Magazine for Radio Amateurs.*

120

Figure 7-10 Architectural Drawing

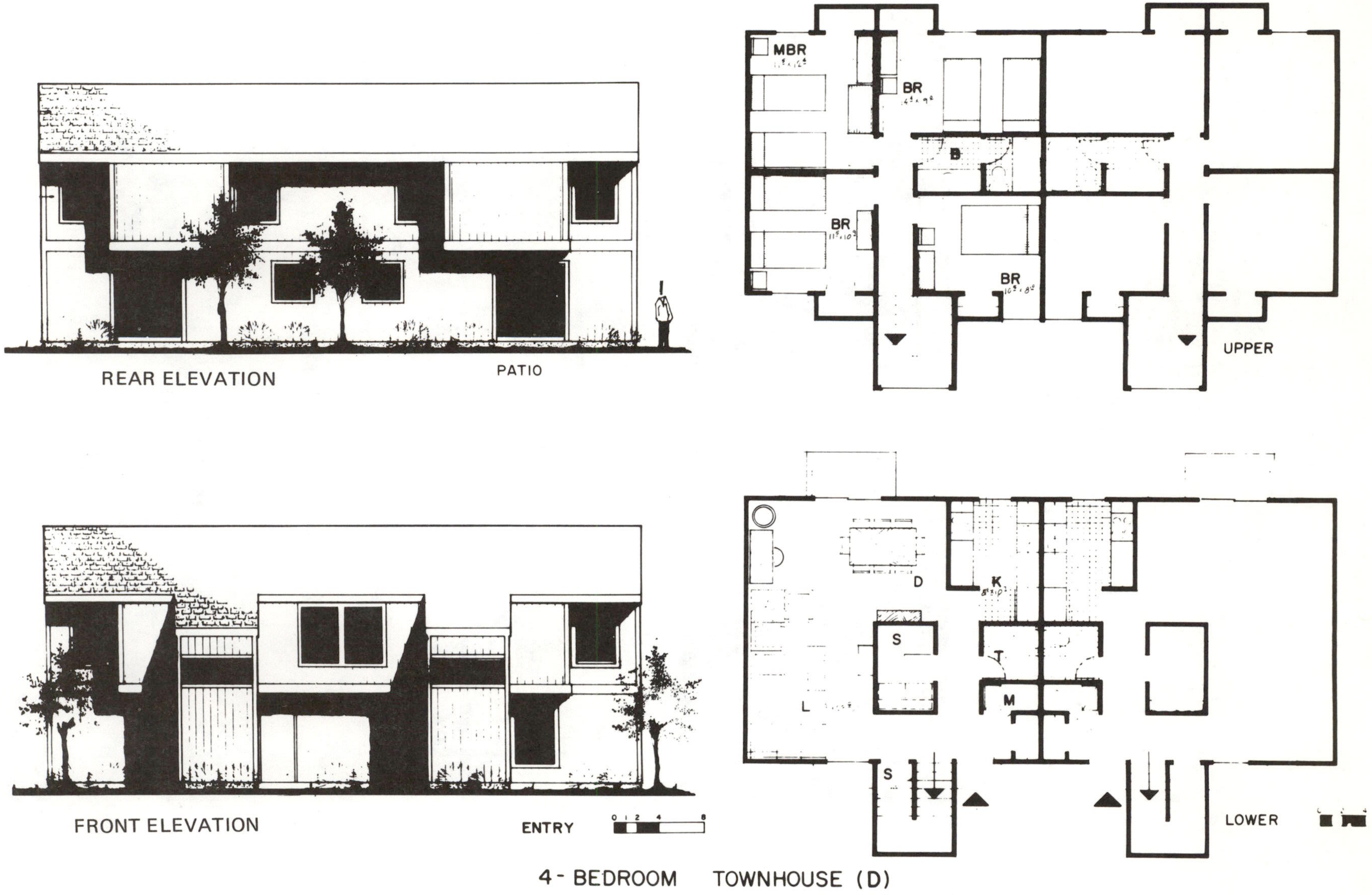

UPPER

MBR

BR

BR

BR

B

LOWER

K

S

D

S

L

T

M

PATIO

REAR ELEVATION

ENTRY

FRONT ELEVATION

4 - BEDROOM   TOWNHOUSE (D)

Courtesy of Mills, Obenchian, Oliver and Webb, Inc., Blacksburg, Va.

Figure 7-11 Structural Steel Drawing

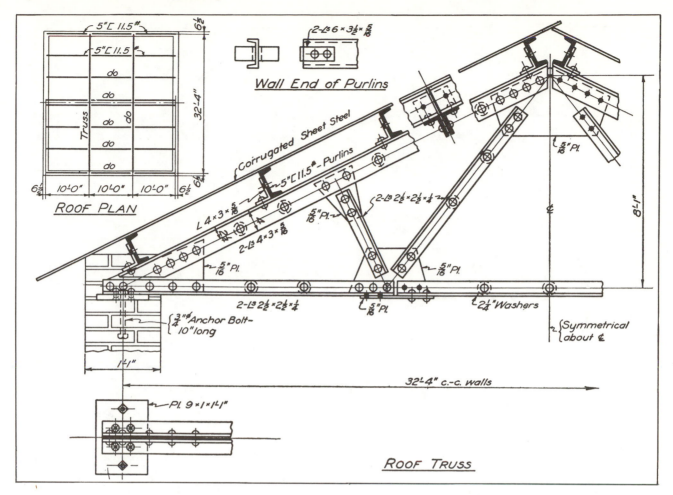

Architectural drawings are usually in the form of plans (top views) and elevations (front and side views). Standard drawing practices are used, but additional features, such as door and window schedules, roof pitches and construction details, are shown to facilitate interpretation. Rendered perspective drawings are often prepared to present a structure in its natural setting. Figure 7-10 shows a simplified plan and rendered elevations.

Structural steel drawings show both detail and assembly information for all steel members. Figure 7-11 is a drawing of a steel truss constructed of angles, plates, washers, and rivets.

Topographic drawings show contours, boundaries, distances, and directions describing a portion of the earth's surface with cultural and man-made features as required. Figure 7-12 shows a typical topographic drawing of a parcel of property.

Figure 7-12 Topographic Drawing

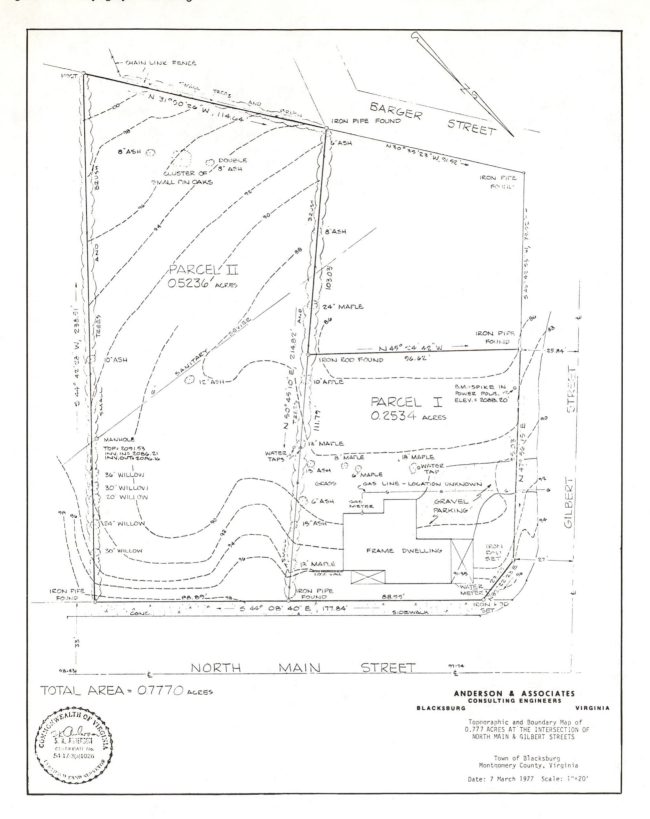

122

Vectors

8

Engineering problems involving mechanisms, structures, space mechanics, or even electrical phenomena can be solved by using a simulation technique involving vectors. The graphical methods outlined in this chapter may be used in solving such problems. Graphical methods are also useful as a means for checking mathematical analysis.

8.1 DEFINITIONS

A *vector quantity* has both magnitude and direction, while a *scalar quantity* has only magnitude. Force, acceleration, and velocity are examples of vector quantities. A *vector* is a directed line segment whose length is proportional to the magnitude of the vector quantity.

Concurrent vectors are vectors whose lines of action intersect at a common point. *Nonconcurrent vectors* do not intersect at a common point. Vectors lying in the same plane are *coplanar*.

A *resultant* is a vector obtained by the addition of two or more vectors and may replace the given vector system. The *equilibrant* is equal and opposite in direction to the resultant. The equilibrant will balance the vector system and cause the system to be in equilibrium.

A *free-body diagram* (FBD) is a drawing that shows the vectors acting on an isolated point or system. A *vector diagram* is a graphical technique for solving vector problems. Any single vector may be replaced by or resolved into a system of two or more vectors, such as a horizontal component and a vertical component.

8.2 SPACE DIAGRAM AND VECTOR DIAGRAM

A space diagram is a drawing of the vector system and is not necessarily drawn to scale. However, the magnitude and direction of each vector quantity are indicated. Figure 8-1 illustrates a space diagram and a vector diagram of three concurrent forces. The resultant R of these three forces is a single vector replacing the given force system.

Figure 8-1

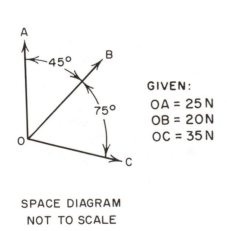

GIVEN:

OA = 25 N
OB = 20 N
OC = 35 N

SPACE DIAGRAM
NOT TO SCALE

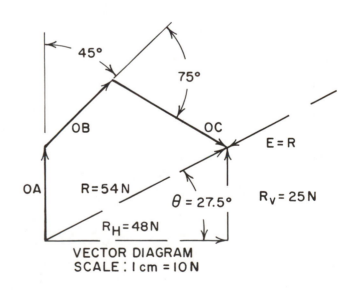

VECTOR DIAGRAM
SCALE : 1 cm = 10 N

The resultant is determined by graphical vector addition. An arbitrary point is selected to begin the vector diagram. *OA* is drawn in the given direction to the given scale. *OB* is added to the tip of *OA* in a similar manner. Continue the diagram by adding *OC* to *OB*. The diagram is completed by drawing the resultant *R* from the starting point to the tip of *OC*. The magnitude of the resultant, the magnitude of the horizontal and vertical components, and the angle the resultant makes with the horizontal may be measured.

A single force of equal magnitude and opposite in direction to the resultant *R* will balance the given forces and will cause the system to be in equilibrium. Such a force is called an equilibrant *E*.

8.3   FREE-BODY DIAGRAM AND BOW'S NOTATION

Consider a weight of 200 N suspended from a structure consisting of a horizontal member and an angular member secured to a wall, as shown in Figure 8-2. In order to design a satisfactory structure, supporting brackets, and other hardware, it is necessary to determine the force in each member.

Figure 8-2

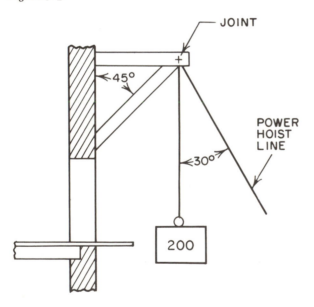

A free-body diagram (FBD) consisting of the structural members and the hoisting line is sketched for the joint. This is a single line drawing of the system, showing each line broken a short distance from the center of the joint. Vectors are identified by *Bow's notation*, using letters adjacent to both sides of the member. Bow's notation consists of letters or numbers placed in each

area between members of the force system. See Figure 8-3. The conventional clockwise system designates the tail of the vector by the first letter and the tip by the second letter.

Figure 8-3

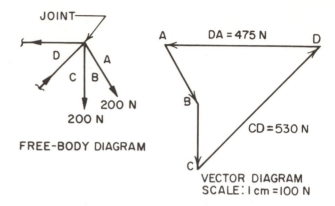

Vector *AB* is drawn first at 30° with the vertical with a magnitude of 200 N. Vector *BC* is then added to vector *AB*. Since force acts along the axis of a structural member, draw a line parallel to the 45° support through point *C* and another line parallel to the horizontal support through point *A*. These lines intersect at point *D*. Arrowheads are added tip to tail to close the vector diagram. The magnitude of vector *CD* and *DA* is measured using the given scale.

Arrowheads are added to complete the FBD. Reference is made to the vector diagram to determine whether the arrowheads point to the joint or away from the joint. The arrowheads in the FBD are placed in the same direction as the direction of the forces in the vector diagram. If the arrowheads are away from the joint, as is the case for vectors *AB*, *BC*, and *DA*, the members are in tension. If the arrowhead is directed toward the joint, as is the case for vector *CD*, the member is in compression. The method just described is extremely useful in determining tensile or compressive stress of members in complex structures.

8.4   BEAM ANALYSIS

A beam secured at one end with a pin and resting on a roller at the other end presents a typical structural design problem. With the beam loaded as indicated in the space diagram in Figure 8-4, it is desired to determine the reactions at the pin and roller.

The pin and roller reactions will be through their respective axes. Further, the reaction at the roller will be through the point of contact between the roller and the beam. This means that the line of action of the

124

roller reaction is through its axis and perpendicular to the beam. The direction of the reaction at the pin is unknown.

Figure 8-4

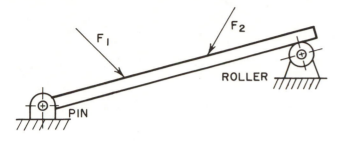

A single line drawing of the space diagram is drawn to scale in Figure 8-5. A partial vector diagram is then drawn using the given vector scale.

Bow's notation is used to identify each force in the space diagram, which is the basis for the *funicular diagram* shown in Figure 8-6. The same notation has been applied to the vector diagram, which is used to develop the *ray diagram*.

An arbitrary pole point *P* is selected and the rays *PA*, *PB*, and *PC* are drawn. A good rule of thumb to use in selecting the location of a pole point is to draw a horizontal line midway between the extreme points on the vector diagram and place the pole point on this horizontal line far enough from the vector diagram so that the extreme rays *PA* and *PC* make an angle of 90° or less.

Figure 8-5

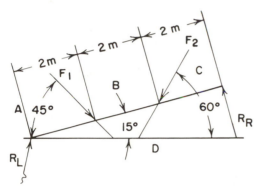

SPACE DIAGRAM
SCALE : 1 : 100

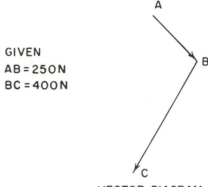

GIVEN
AB = 250N
BC = 400N

VECTOR DIAGRAM
SCALE : 1cm = 200N

Figure 8-6

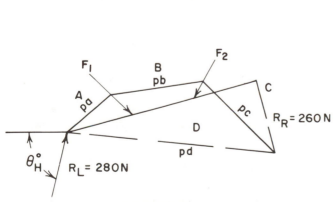

FUNICULAR DIAGRAM
SCALE : 1 : 100

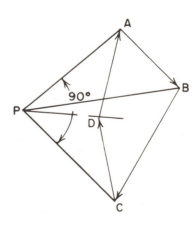

RAY DIAGRAM
SCALE : 1cm = 200N

The funicular diagram must be started at the pin because the pin reaction passes through the axis of the pin but its direction is not known.

- Draw a line parallel to the ray *PA* from the pin to the force *AB* of the space diagram.
- Draw a line parallel to the ray *PB* in area *B* connecting to *pa*.
- Draw a parallel to ray *PC* in the area *C*.
- Close the polygon by drawing *pd* with a dashed line. *PD* is the ray missing from the ray diagram.
- Draw a parallel to *pd* through the pole point *P*.
- Draw a parallel to the roller reaction through point *C* on the vector diagram. Point *D* is located where these last two lines intersect.
- Join *D* to *A* to finish the vector diagram. The magnitude of *CD* and *DA* can now be determined.
- The pin reaction is then drawn on the space diagram parallel to the vector *DA*. The angle the pin reaction makes with the horizontal is measured as 76°.

8.5 TRUSS ANALYSIS

A simple truss is loaded as shown in Figure 8-7. To determine the reactions at the roller and the pin, and the magnitude and nature (tension or compression) of the forces in the structural members, first apply Bow's notation to the outside areas between the known forces as well as to the inside areas between the structural members.

Figure 8-7

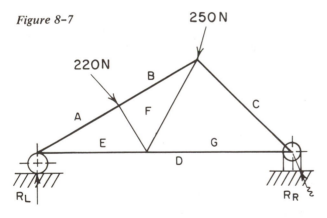

The funicular diagram and the ray diagram in Figure 8-8 are drawn by following the same procedures that were used for the beam analysis in Section 8.4. Notice that the lines of action of the forces *AB* and *BC* had to be extended in order to terminate the sides of the funicular diagram. The magnitude and directions of the reactions are now known.

In Figure 8-9, the funicular portion of the space diagram and the ray portion of the vector diagram are omitted for the sake of clarity. In practice, construction should never be erased from the solution.

126

Figure 8-8

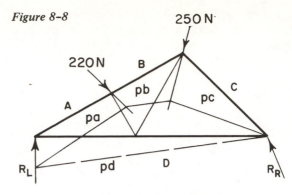

FUNICULAR DIAGRAM
SCALE: 1:100

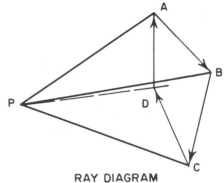

RAY DIAGRAM
SCALE: 1 cm = 100N

Circles drawn around the junctures of the structural members and the applied forces isolate each joint as a free body from the remainder of the structure. An inspection of joint 1 reveals that the force in members *AE* and *ED* is unknown and that the reaction at the roller *DA* is known. A similar inspection of joint 4 reveals that the force in members *DG* and *GC* is unknown and that the reaction at the pin *CD* is known. An inspection of joint 2 reveals that the force in members *BF*, *FE*, and *EA* is unknown and that the force *AB* is known. In joint 3, the force in members *CG*, *GF*, and *FB* is unknown and force *BC* is known. In joint 5, the force in members *DE*, *EF*, *FG*, and *GD* is unknown. Since joints 2, 3, and 5 have three or more unknown forces, no immediate evaluation can be made for the structural members in these joints. Joints 1 and 4 each have only two unknown forces that can be determined graphically. Then the number of unknown forces in each of the other joints will be reduced to no more than two and evaluations can be made.

The composite vector diagram is started by drawing a parallel to structural member *AE* through point *A*. Then draw a parallel to *ED* through point *D*. The intersection of these two lines fixes point *E* in the composite vector diagram. The magnitudes of *AE* and *ED* can then be scaled. The roller reaction *DA* is vertically upward on

Figure 8-9

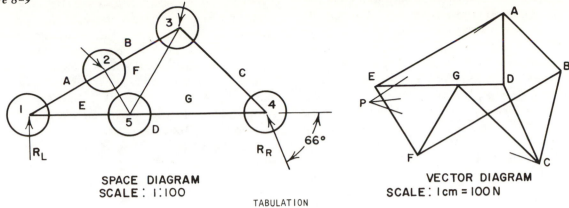

SPACE DIAGRAM
SCALE : 1:100

VECTOR DIAGRAM
SCALE : 1cm = 100 N

TABULATION

Forces in Newtons, N; C indicates Compression, T indicates Tension

Known	Scaled from Vector Diagram	
AB = 220	AE = 380 C	GC = 300 C
BC = 250	ED = 330 T	GF = 215 T
CD = R_R = 230	BF = 440 C	DG = 120 T
DA = R_L = 190	FE = 215 C	

the space diagram and also on the composite vector diagram. An examination of the FBD of joint 1 indicates that the direction of the force in *AE* is toward the joint, in compression, and the force in *ED* is away from the joint, in tension.

Through point *C*, draw a parallel to *GC* as far as the horizontal line *DE* because structural member *DG* is also horizontal. The pin reaction is in the direction from *C* to *D*. Therefore, *DG* is away from the joint, in tension, and *GC* is toward the joint, in compression. The composite vector diagram for joints 2, 3, and 5 is determined in a similar manner. The results are tabulated.

8.6 CONCURRENT NONCOPLANAR VECTORS

Frequently, concurrent force systems involving three or more forces do not lie in the same plane. In such cases, the space diagram and the vector diagram will be multi-view drawings and the vector representing a force may or may not be true length in any of the given views. In laying out the vector diagram for Figure 8-10, the directions of the vectors are obtained from the space diagram. The vectors are scaled using the vector scale and the tabulation of forces.

Draw V_1 first. Any line segment 0-1 is used to start

Figure 8-10

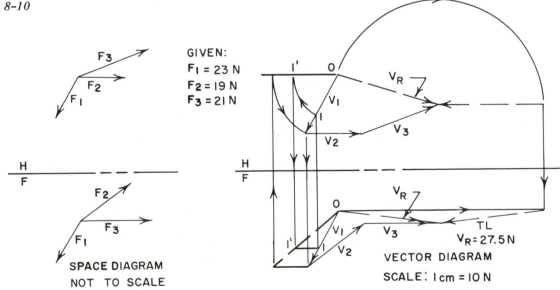

GIVEN:
F_1 = 23 N
F_2 = 19 N
F_3 = 21 N

SPACE DIAGRAM
NOT TO SCALE

VECTOR DIAGRAM
SCALE : 1cm = 10 N

V_R = 27.5 N

127

the construction. Then rotate 0–1 until it is parallel to the frontal plane. The same segment 0–1 is rotated in the front view to 0–1'. The true length of V_1 is laid off in the front view and then it is counter-rotated back to its original position in the front and top views. Draw V_2 true length in the front view and find the top view by projection. Draw V_3 true length in the top view and find the front view by projection. Now, V_R can be added to the vector diagram. The true length of V_R is found by rotation.

The resolution of the component forces in a three member truss supporting a load is shown in Figure 8–11. Two of the members, 0–3 and 0–4, lie in a plane perpendicular to the frontal plane and their projections coincide in the front view. This relationship simplifies the solution and must be obtained in an auxiliary view if not shown in a principal view of the space diagram.

The vector diagram is started in both views with vector V_1 representing the 40 N load. The diagram is completed by drawing vectors parallel to the corresponding lines in the space diagram.

The magnitude of the forces is scaled in the true length views of the vectors.

Figure 8–11

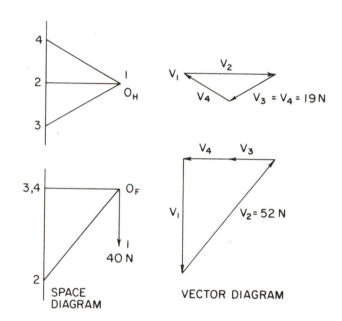

Charts and Graphs

9

Engineers, scientists, and businessmen are constantly faced with the problem of collecting, evaluating, and presenting data and information. Many occasions arise when such information can best be presented in a condensed and easily understandable form. The capability to accomplish this through judicious design and use of charts and graphs is a skill which is invaluable in professional work.

9.1 BAR CHARTS

Bar charts generally present numerical information through the use of "bars" drawn to a convenient scale. The bars may be horizontal or vertical, contiguous or separated. Bar charts are easily prepared and provide excellent visual recognition of information. Figure 9–1 contains typical bar charts.

Figure 9–1 Bar Charts

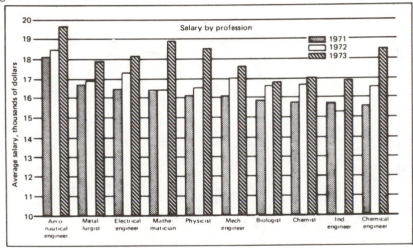

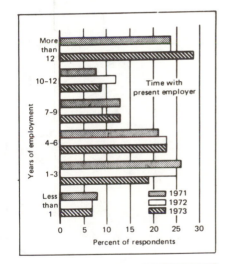

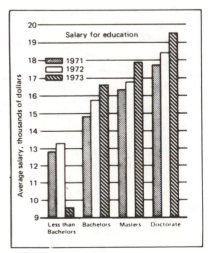

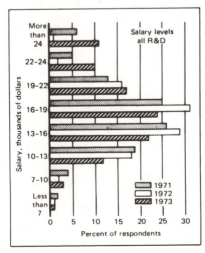

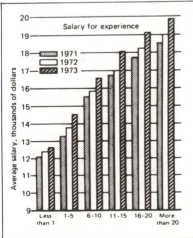

INDUSTRIAL RESEARCH – MAR 1973

9.2 PICTORIAL CHARTS

Pictorial charts are basically artistic adaptations of bar charts. They depict relative quantities by the size of easily identified picture symbols. Care must be exercised in designing and reading such charts to avoid erroneous impressions. See Figure 9-2.[1] Comparison of Figures 9-2 and 9-3 indicates a dramatic increase in the income of engineers during a decade.

Figure 9–2   Pictorial Chart

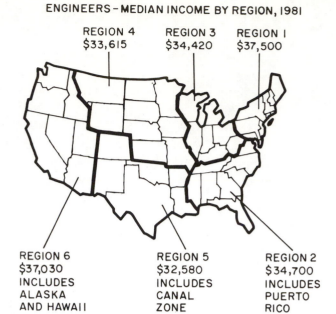

Figure 9–3   Area Chart

ENGINEERS – MEDIAN INCOME BY REGION, 1981

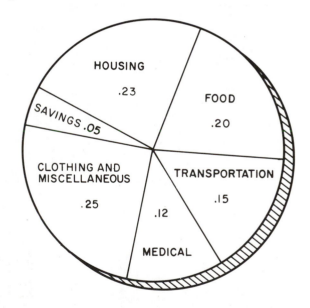

Figure 9–4   Pie Chart

A TYPICAL FAMILY DOLLAR

9.3   AREA CHARTS

Area charts, such as the common "pie chart," readily depict information as parts of a whole. They are most useful in showing percentages or fractional breakouts. See Figures 9-3 and 9–4.

9.4   ORGANIZATION CHARTS

Organization charts are an indispensable management tool to illustrate organizational structure as well as lines of responsibility, coordination, and communication. A typical simplified chart for a small industrial organization might appear as in Figure 9–5.

9.5   PERT CHARTS

The letters PERT stand for "*P*rogram *E*valuation and *R*eview *T*echnique." A *PERT chart* graphically portrays activities and events that must be accomplished for

[1]Figures 9-2 and 9-17 are reprinted from *Civil Engineering,* the official monthly publication of the American Society of Civil Engineers. Data for Figure 9-3 are extrapolated from *The Virginia Engineer,* December 1981.

Figure 9-5   Organization Chart

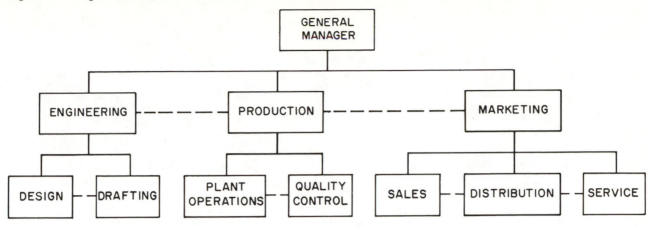

completion of a project. The events are arranged in proper time sequence and connected by activity lines to develop a network of events. From this network a *critical path* may be determined. This path is the one that requires the longest time from beginning to completion of a project, hence it controls the schedule. This technique, in various forms, has proven to be a popular and efficient management tool. See Figure 9-6.

Figure 9-6   PERT Diagram

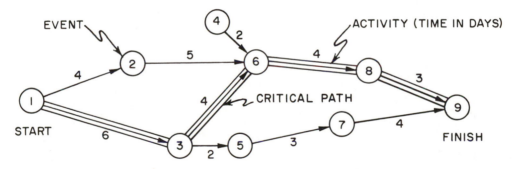

9.6   GRAPHING PROCEDURE— LINE GRAPHS

1. Select an appropriate grid that is best suited to the data. There are many types of commercially prepared graph paper, or special grids may be ruled on plain paper for a particular problem. Types of grids are discussed later.
2. Select the location of the axes so that all data can be plotted.
3. Scale each axis to fit the available space.
4. Identify, label, and graduate each axis so that data may be easily interpreted. Normally, the independent variable is associated with the horizontal or X-axis.
5. Plot the data points using easily identified symbols. This is particularly important if more than one curve is involved on the graph.
6. Draw the curve. For a continuous function, a smooth curve is first sketched through the data points to obtain a "best fit" and then drawn with the aid of a French curve and straightedge. Discontinuous data points are connected by straight lines. Lines should not continue through the plotted point symbols.
7. Letter the title of the graph and any necessary notes. Do not clutter the graph. All lettering should read from the bottom and right side of the graph. See Figures 9-7 and 9-8.

Figure 9-7 Line Graph

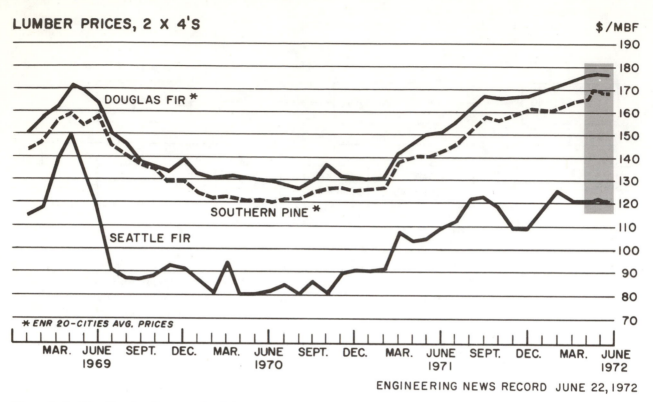

LUMBER PRICES, 2 X 4'S

$/MBF

DOUGLAS FIR *

SOUTHERN PINE *

SEATTLE FIR

* ENR 20-CITIES AVG. PRICES

MAR. JUNE SEPT. DEC. | MAR. JUNE SEPT. DEC. | MAR. JUNE SEPT. DEC. | MAR. JUNE
1969 | 1970 | 1971 | 1972

ENGINEERING NEWS RECORD JUNE 22, 1972

Figure 9-8 Line Graphs–Rectangular Grid

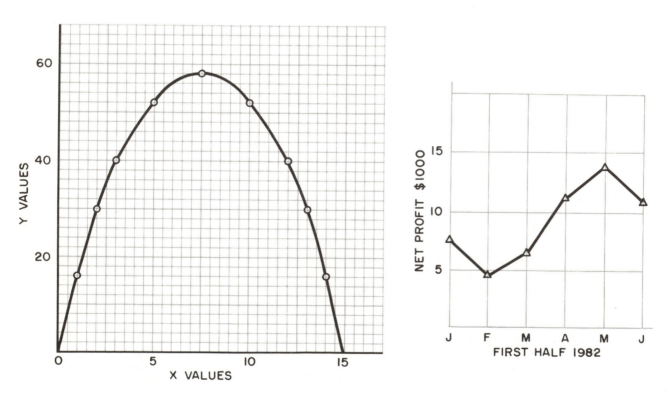

9.7 GRIDS

Rectangular grids. *Rectangular grids* are most commonly used to plot line graphs, either straight lines or curves. The horizontal and vertical lines comprising the grid may or may not have the same spacing. See the examples in Figure 9–8.

Logarithmic grids. *Logarithmic grids* are special rectangular grids graduated into logarithmic divisions along the ordinate, abscissa, or both. If both scales are logarithmic, it is a logarithmic grid. If one scale is logarithmic (usually the vertical scale) and the other arithmetic, it is a semilogarithmic grid.

Commercially printed logarithmic graph paper is available in many forms and cycle lengths. There is no zero point on a logarithmic scale, and negative values cannot be plotted. Repetition of cycles allows plotting data ranging from very small to very large numbers, as each cycle is raised by a factor of ten.

The graphs of products, quotients, and variables with exponents plot as straight lines on a logarithmic grid. The equation $XY = 10$ plots as a curve on a rectangular grid and as a straight line on a logarithmic grid. See Figure 9–9.

Semilogarithmic grids. The *semilogarithmic grid* is sometimes called a rate-of-change grid because functions whose values are in geometric progression plot as straight lines on this grid. The slope of the line then shows the rate of change or relative change rather than the absolute change in values. Curves of equations with a variable exponent plot as straight lines on semilogarithmic grids. See Figure 9–10.

Figure 9–9

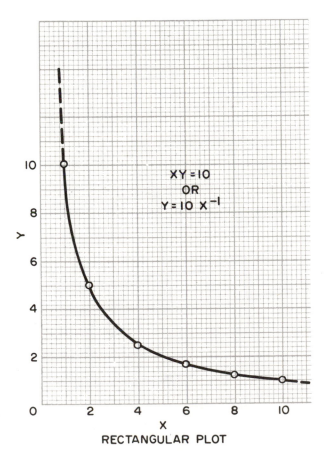

RECTANGULAR PLOT

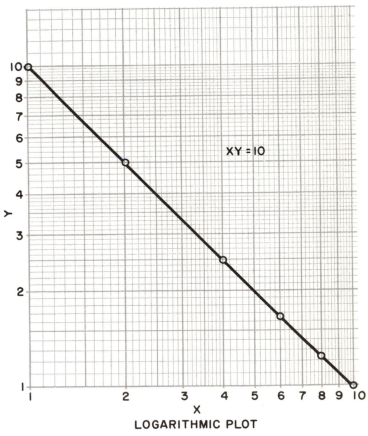

LOGARITHMIC PLOT

Figure 9-10

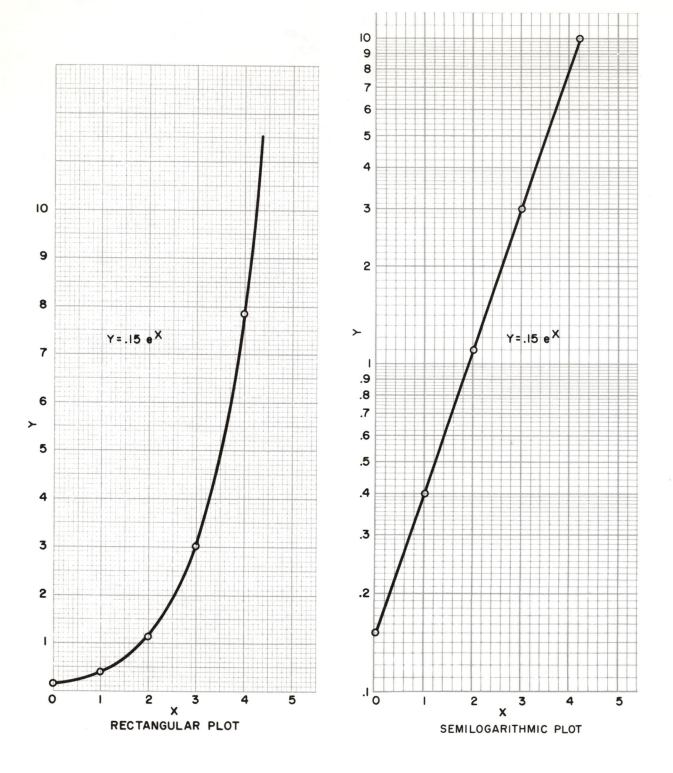

134

Polar grids. *Polar grids* consist of equally spaced concentric circles and equally spaced radius lines from a common center known as the *pole*. This grid allows the continuous plot of two variables, one angular and the other linear. The polar graphs in Figure 9-11 show the horizontal field strength pattern of two antennas, one of which is bidirectional and the other a corner reflector.

Figure 9-11 Polar Graphs

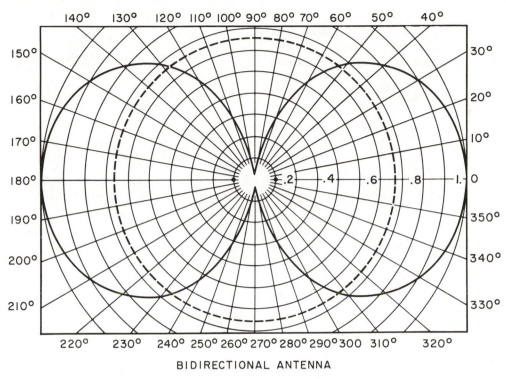

BIDIRECTIONAL ANTENNA

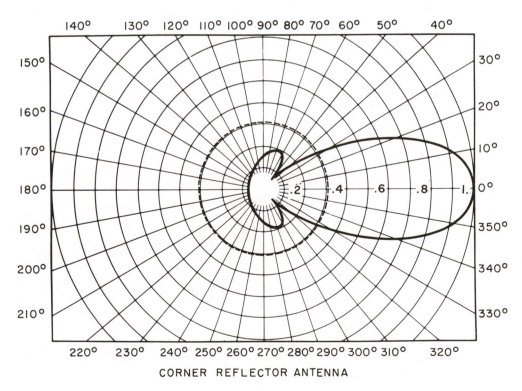

CORNER REFLECTOR ANTENNA

135

9.8 STRAIGHT-LINE, POWER, AND EXPONENTIAL CURVES

The curves most frequently encountered in engineering are the straight-line, power, and exponential curves. If it can be determined which of these curves most closely approximates plotted data, then the mathematical relationship of the variables (the *equation of the curve*) may be obtained. Such equations, derived from experimental data, are known as *empirical equations.*

The simplest curve is the *straight line.* If plotted data on a rectangular arithmetic grid approximate a straight line, the equation of the line may be determined from the general equation

$$y = mx + b$$

where

m = slope of the line
b = value of y where $x = 0$ (the y-intercept)

Consider the plotted data in the Figure 9-12. A "best-fit" line has been drawn through the plotted points. The slope of the line is easily scaled to give the value of m and the y-intercept, b, is also easily determined.

The family of *power curves* is related to curves derived from the intersection of a right circular cone and a plane. See Figure 9-13. These are known as conic sections and include the circle, ellipse, parabola, and hyperbola. In general, if the cutting plane makes an angle θ with the axis of a cone whose generating angle is β, the section will be:

a circle, when $\theta = 90°$
an ellipse when $90° > \theta > \beta$
a parabola, when $\theta = \beta$
a hyperbola, when $0° < \theta < \beta$

If the graph of empirical data on rectangular grid is a curve, it may be a power curve of the form

$$y = bx^m$$

which will either pass through the origin (a parabola; m is positive and > 1) or be asymptotic to the axes (a hyperbola, m is negative).

Figure 9-12

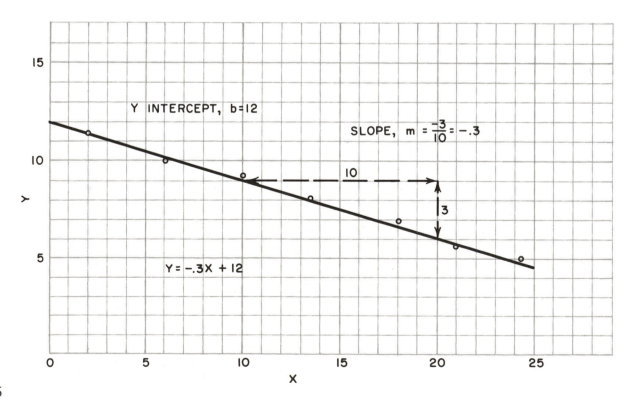

Figure 9–13 Conic Sections

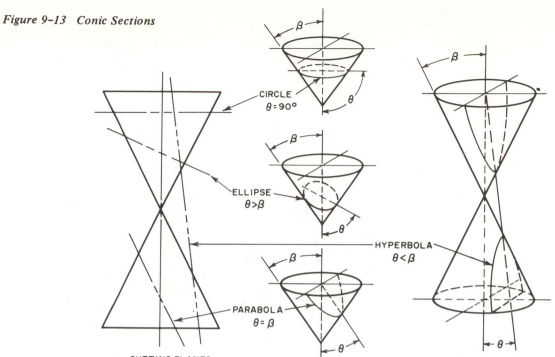

β
CIRCLE
θ=90°
θ
ELLIPSE
θ>β
HYPERBOLA
θ<β
PARABOLA
θ=β
CUTTING PLANES

Figure 9–14 Power Curve

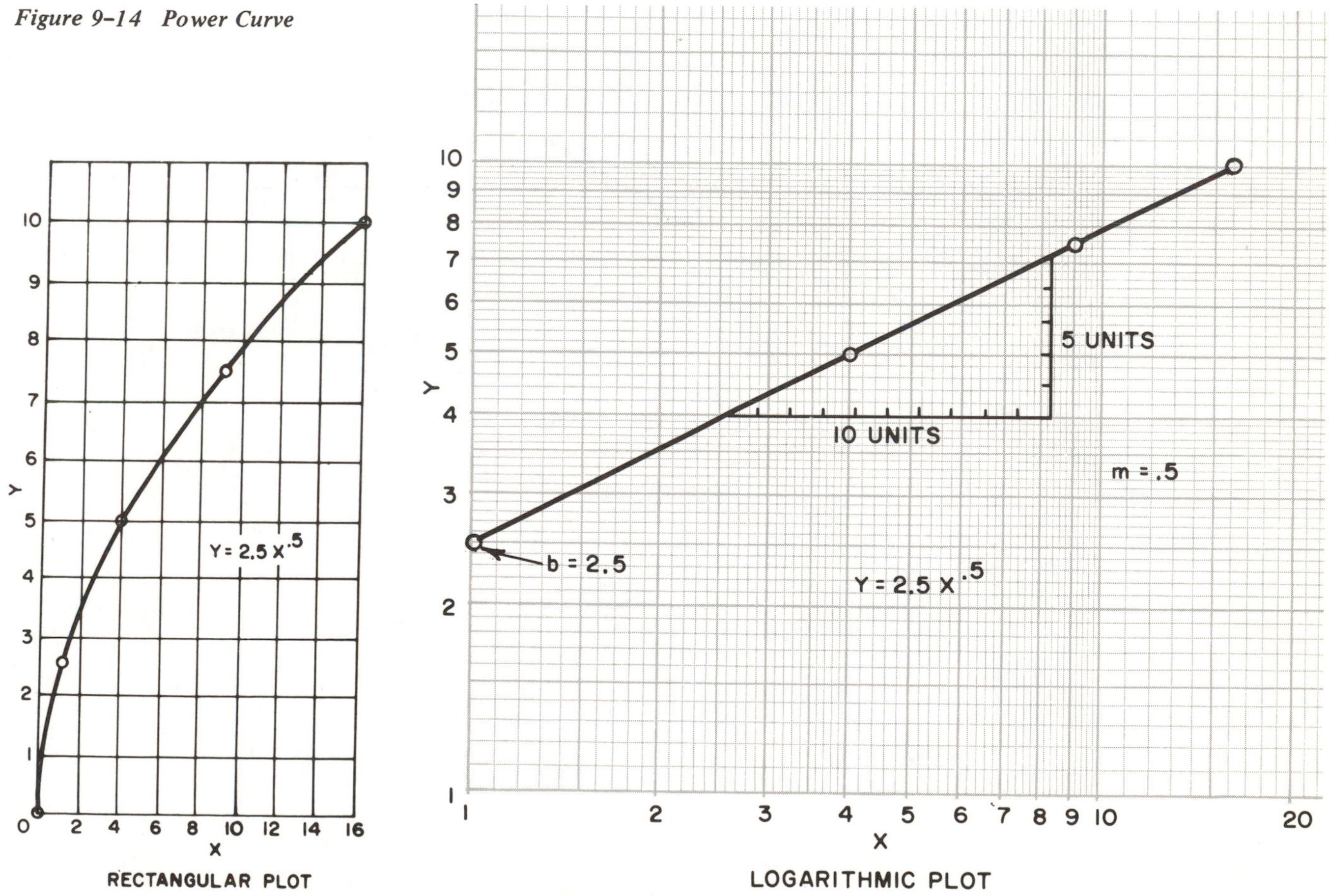

Y
X
0 2 4 6 8 10 12 14 16
1 2 3 4 5 6 7 8 9 10
Y = 2.5 X^.5
RECTANGULAR PLOT
5 UNITS
10 UNITS
m = .5
b = 2.5
Y = 2.5 X^.5
1 2 3 4 5 6 7 8 9 10 20
LOGARITHMIC PLOT

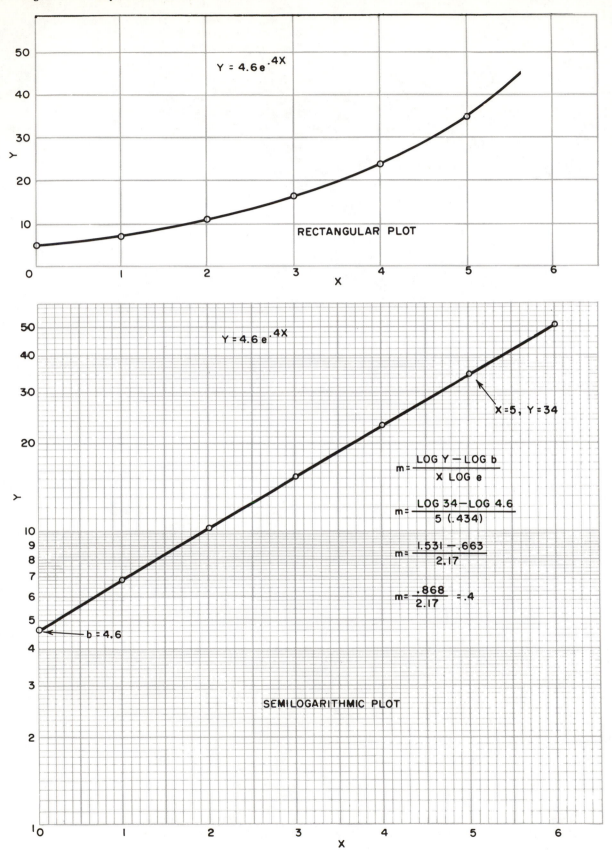

Figure 9-15 Exponential Curve

$$Y = 4.6\,e^{.4X}$$

RECTANGULAR PLOT

$$Y = 4.6\,e^{.4X}$$

X=5 , Y=34

$$m = \frac{LOG\ Y - LOG\ b}{X\ LOG\ e}$$

$$m = \frac{LOG\ 34 - LOG\ 4.6}{5\ (.434)}$$

$$m = \frac{1.531 - .663}{2.17}$$

$$m = \frac{.868}{2.17} = .4$$

b = 4.6

SEMILOGARITHMIC PLOT

Power curves plot as straight lines on a logarithmic grid since the preceding equation may be expressed as

$$\log y = m \log x + \log b$$

which is similar to the straight-line equation, $y = mx + b$, except that here logarithms are involved.

The slope, m, of the straight line plotted on the logarithmic grid may be directly scaled and the value of b read directly. Note that the value of b must be read at $x = 1$, where $\log x = 0$. See Figure 9–14.

If the graph of a curve intersects only one axis, the curve may be of the exponential type. *Exponential curves* plot as straight lines on a semilogarithmic grid and take the form

$$y = be^{mx} \text{ (most common)} \qquad (1)$$

$$\text{or}$$

$$y = bm_1{}^x \qquad (2)$$

where b is the y intercept at $x = 0$. The independent variable appears as the exponent. Since the semilogarithmic grid is a combined arithmetic and logarithmic grid, the value of m or m_1 *must be calculated* by substituting values of x, y, and b taken from the straight line plot.

In the case of equation (1),

$$y = be^{mx}$$

Then $\log y = \log b + mx \log e$ and

$$m = \frac{\log y - \log b}{x \log e} \qquad (\textit{Note: } \log e = 0.434)$$

In the case of equation (2),

$$y = bm_1{}^x$$

Then $\log y = \log b + x \log m_1$ and

$$\log m_1 = \frac{\log y - \log b}{x}$$

m_1 is then the antilog of $\log m_1$. From the above it is apparent that

$$e^m = m_1 \qquad \text{or} \qquad m \log e = \log m_1$$

The equations derived will be the same even though they are of a slightly different form. See Figure 9–15.

9.9 RESOLUTION OF EMPIRICAL DATA, CURVE FITTING

From the preceding discussion it is evident that the empirical equation for experimental data may easily be determined from the graph of the data. The steps to be followed are:

1. Plot the data on a rectangular grid and draw the best-fit curve.
2. If the original curve approximates a straight line, derive the equation in the form

$$y = mx + b$$

3. If the curve appears to be a parabola or hyperbola, replot the curve on logarithmic grid. If the plot is a straight line, derive the equation in the form

$$y = bx^m$$

4. If the original curve crosses an axis or if the logarithmic grid plot is also a curve, plot the original curve on semilogarithmic grid. If this plot is a straight line, derive the equation in the form

$$y = be^{mx}$$

In all cases, care must be exercised in accurately plotting the given data and drawing the curves.

9.10 NOMOGRAPHY

Nomography, in the mathematical sense, is the art of graphically representing numerical relations. A *nomograph* is a chart that enables the user to determine an unknown value or values of variables when one or more values of the numerical relationship are known.

While nomographs may be designed in a variety of ways, some quite complex, the most frequently used form is the *alignment chart.* This is a chart consisting of two or more graduated scales arranged for the solution of a single equation within specified limits.

Examples of various types of alignment charts are the adjacent scale, parallel scale, and "N"-type charts shown in Figure 9–16. Values satisfying the single equation are read at the intersections of a straight line with the scales. The straight line is called an *isopleth.*

These basic forms of alignment charts may be combined to handle additional variables (Figure 9–17) by creating a *dummy* or *turning scale* for the connection of isopleths crossing each segment of the combination chart.

Figure 9-16   Alignment Charts

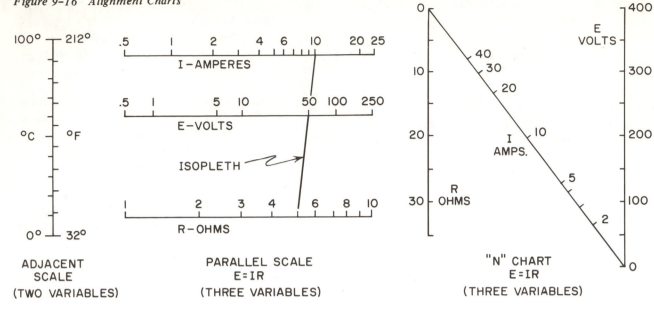

ADJACENT
SCALE
(TWO VARIABLES)

PARALLEL SCALE
E=IR
(THREE VARIABLES)

"N" CHART
E=IR
(THREE VARIABLES)

Figure 9-17   Alignment Chart—Four Variables

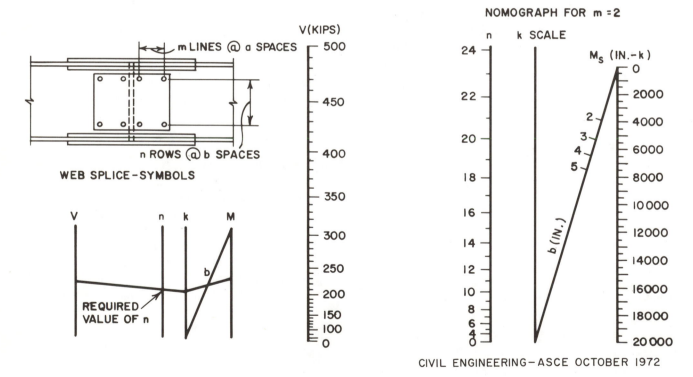

WEB SPLICE-SYMBOLS

NOMOGRAPH FOR m = 2

CIVIL ENGINEERING—ASCE OCTOBER 1972

Nomographs allow a quick and easy solution for fixed numerical relationships. As such, they are very useful in industry and often in the solution of engineering problems. It is not the intent of this text to develop the methodology for the construction of nomographs. A complete explanation of these techniques may be found in any of the many available nomography texts.

Graphical Calculus

10

In the preceding chapter, techniques were developed for the plotting of data and the determination of the equation of best-fit curves. When the equation of a curve is known, relationships among the variables may be obtained by the traditional methods of the calculus. Frequently, however, when data cannot be expressed in the form of a standard equation or when only one rate or integrated effect of the variables is needed, it is easier to solve for the desired results using the techniques of graphical calculus.

10.1 GRAPHICAL REPRESENTATION OF THE DERIVATIVE

Differentiation is the determination of the rate of change of one variable with respect to the other. The rate at which one variable $y = f(x)$ varies with respect to another, x, is defined mathematically as

$$\operatorname*{limit}_{\Delta x \to 0} \frac{f(x + \Delta x) - f(x)}{\Delta x} = \operatorname*{limit}_{\Delta x \to 0} \frac{\Delta y}{\Delta x} = \frac{dy}{dx}$$

Here dy and dx are infinitesimally small increments of the variables y and x, respectively, and their ratio dy/dx is called the *derivative of the function*. Referring to Figure 10-1, we see that $\Delta y/\Delta x = \tan \gamma$, and that as $\Delta x \to 0$, $\gamma \to \alpha$ in the limit. But $\tan \alpha = dy/dx =$ the rate of change of y with respect to x. Geometrically, this ratio dy/dx may be interpreted as the *slope* of a tangent to the plotted curve at x. Since dy and dx are actually infinitesimally small, they cannot be measured. However, $dy/dx = \delta y/\delta x$, and both δy and δx of Figure 10-1 can be measured to any convenient scale. Alternatively, α may be measured in degrees if, and only if, x and y are plotted to the same scale, and then

$$\tan \alpha = dy/dx = \text{slope of the tangent at any point on the curve}$$

Thus, when $y = f(x)$ is an unknown mathematical relationship, its derivative at any specific value of x may be determined by (1) plotting experimental data, (2) drawing a smooth curve through the plotted data points, (3) constructing a tangent to the curve wherever the derivative is desired, and (4) measuring either δy and δx or α to find dy/dx.

Figure 10-1

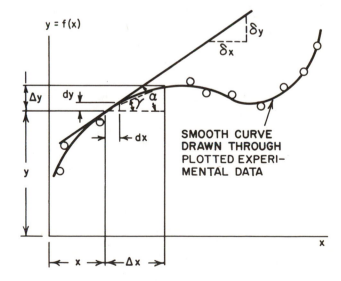

SMOOTH CURVE DRAWN THROUGH PLOTTED EXPERIMENTAL DATA

10.2 TANGENT LINE CONSTRUCTION

A line tangent to a curve may be constructed most easily by using a reflecting surface such as a stainless steel or glass mirror. The reflecting surface is placed normal to the curve at the desired point of tangency and the surface adjusted until the mirror image and the curve to its front appear as a smooth, unbroken line. The tangent line will then be perpendicular to a line drawn along the reflecting surface. The slope of the tangent at the desired point may then be evaluated by measurement.

Example. A cast-iron tensile specimen was tested to fracture. The load was applied in increments, elongations measured, these data recorded after converting to unit stress and strain, and plotted as shown in Figure 10-2. Determine the tangent modulus of elasticity E_t at a stress of 100 megapascals (MPa).
Note: One pascal equals a pressure of one newton per square meter.

Figure 10-2

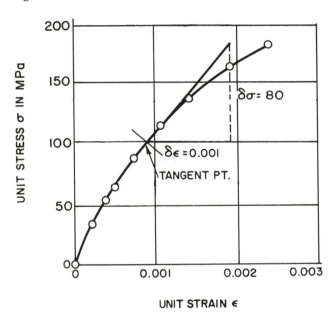

UNIT STRAIN ε

Solution. Construct a tangent to the plotted curve at $\sigma = 100$ MPa.

$$\delta\sigma = 80 \text{ MPa}$$

$$\delta\epsilon = 0.001$$

Then calculate

$$E_t = \delta\sigma/\delta\epsilon = 80/0.001 = 8 \times 10^4 \text{ MPa}$$

Note: Stress is defined as force per unit area, or $\sigma = dF/dA$. Strain is defined as deformation per unit length, or $\epsilon = dL/L$. *Modulus of elasticity* is defined as the ratio of stress to strain or $E = d\sigma/d\epsilon$.

10.3 THE SLOPE LAW AND DERIVED CURVES

Differentiation is the process of determining the slope of a curve at any point. When specific values are determined for the derivative at a series of points, these derivatives may be plotted to generate a curve, a *derived curve*. The slope of the tangent to a curve at a given point is equal to the ordinate of the corresponding point on the derivative curve. This is known as the *slope law*. The technique for plotting the first—and succeeding—derivative curve is illustrated in the following example.

Figure 10-3

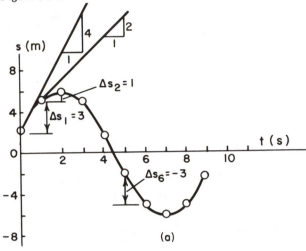

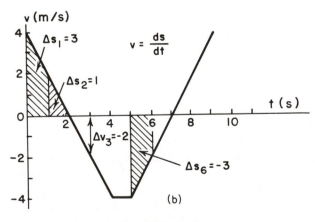

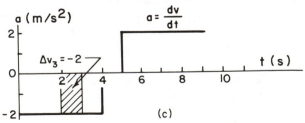

142

Example. A particle is observed to start 2 m to the right of an origin and move along a straight line so that the measured magnitude of its displacement, s, from that origin varies with the measured time, t, as shown in Figure 10-3(a). Plot the variation of the particle's velocity and acceleration.

Solution. Since velocity (v) is defined as the rate of change of displacement with respect to time, and acceleration (a) as the rate of change of velocity with respect to time, we need to evaluate the first derivative of s, $v = ds/dt$, and in turn the first derivative of v, $a = dv/dt$, at a series of selected abscissae, and to plot these results. Figures 10-3(b) and (c) illustrate the results. It should be noted that the particle first moves to the right but with decreasing velocity until it is 6 m from the origin, then speeds up while moving to the left, and then slows down until it stops 6 m to the left of the origin, and then starts moving to the right again.

10.4 THE AREA LAW

The *area law* states that the difference in length of any two ordinates of a continuous curve is equal to the area under the derived curve between the corresponding abscissae. Conversely, the area under a curve between two abscissae is equal to the difference in ordinate of the next higher curve over the same interval. Use of the area law provides a valuable check on the construction of derivative curves using the tangent method and may itself be used to plot a derivative curve.

A derived (v vs. t) curve has been developed in Figure 10-4 from a base (s vs. t) curve. The base curve has been divided into 1-second increments. The difference in ordinate between 0 and 1 second is 2 m. This value is then plotted for the derived curve at the midpoint of the time increment (0.5 s).

Note that the rectangle 0-2-1'-1 has an area of 2 m/s × 1 s = 2 m, the value of the ordinate difference of the s vs. t curve across this increment. The curve itself is then drawn so that the areas "outside" and "inside" the curve are approximately equal (crosshatched areas). This procedure is continued throughout the length of the curve. Note that the ordinate of the derived curve is zero where the slope of the base curve is zero.

10.5 GRAPHICAL INTEGRATION

Graphical integration is the reverse of graphical differentiation. Integration determines the *total change* in the relationship of variables, while differentiation determines the *rate of change*. Integration involves a *summation* of the relationships and may be thought of as the process of determining the area under a given curve between specified limits.

Figure 10-4

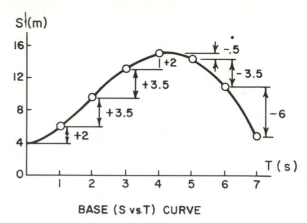

BASE (S vs. T) CURVE

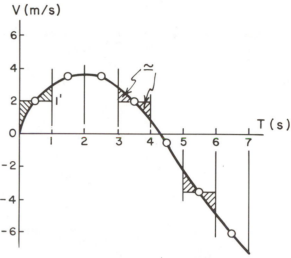

DERIVATIVE (V vs. T) CURVE

The graphical and semigraphical methods of integration are most useful when the equation of a function is unknown or difficult to obtain, as is frequently the case with experimental data.

The integral curve of a function may be derived through use of the area law, which may be restated as follows: The incremental area under a continuous curve is equal to the difference in ordinates of the integral curve between the same limits, provided that the increment is sufficiently small. The proof of this follows from the definition of a derivative. Let

$$v = \frac{ds}{dt} = \text{the slope of the } s \text{ vs. } t \text{ curve}$$

Then

$ds = v\,dt = $ the *increment* of area under the v vs. t curve

or

$$\int_{s_a}^{s_b} ds = \int_{t_a}^{t_b} v\,dt = \begin{array}{l} \text{the } \textit{area} \text{ under the } v \text{ vs. } t \text{ curve} \\ \text{between } t_a \text{ and } t_b \end{array}$$

143

so the *change* in s (or $\Delta s = s_b - s_a$) = the area under the v vs. t curve between t_a and t_b. Both of the integrals in the preceding equation are *definite integrals*. The equation may be modified so that

$$s_b - s_a = \int_{t_a}^{t_b} v\, dt \quad \text{or} \quad s_b = \int_{t_b}^{t_a} v\, dt + s_a$$

or, more generally,

$$s = \int v\, dt + C$$

where C is a constant of integration and $\int v\, dt$ is called the *indefinite* integral. Here C is an arbitrary constant and may have any value. It defines one particular curve (s) of a whole family of such curves.

Example. A particle moves along a straight line so that its velocity varies with time as shown in Figure 10-3(b). Plot the variation of the particle's displacement if it starts ($t = 0$) at a point 2 m to the right of the origin.

Solution.

Given $\qquad v = \dfrac{ds}{dt}$

$$ds = v\, dt$$

$$\Delta s = \int_{t_a}^{t_b} v\, dt = \begin{array}{l} \text{the incremental area} \\ \text{under the } V \text{ vs. } T \text{ curve} \end{array}$$

Let the increments in $\Delta t = 1$ s. Then for $0 < t < 1$,

$$\Delta s_1 = \int_0^1 v\, dt = \frac{4 + 2}{2}\,(1) = 3 \text{ m}$$

Similarly,

$$\Delta s_2 = \int_1^2 v\, dt = \frac{2 + 0}{2}\,(1) = 1 \text{ m}$$

Now, starting at $s = +2$ m, a series of Δs increments may be plotted at each succeeding second, generating the s vs. t curve shown in Figure 10-3(a).

10.6 STRING POLYGON METHOD— GRAPHICAL INTEGRATION

The string polygon method of integration involves the summation of the areas of a series of rectangles closely approximating incremental areas under a given curve. The chosen increments need not be of equal width. Smaller increments should be used where the curve has a steep slope.

This method is illustrated in Figure 10-5. First determine the approximate value of the area under the curve, using the given X and Y units. This provides the range for the ordinate scale of the integral curve.

Next, divide the curve into successive areas by vertical lines and approximate the areas by drawing horizontal lines through the curve so that the small area of the rectangles *outside* the curve is approximately equal to the small area above or below the horizontal lines *inside* the curve.

Select a "pole point" on an extension of the X-axis so that the pole point distance, PO, is some multiple of a unit of distance on the X-axis. The greater the multiple used, the flatter the integral curve to be determined.

Project the tops of the approximate rectangles to the Y-axis, or to a line parallel to the Y-axis. From the pole point, draw lines to the intersection of these projections with the Y-axis or the parallel line. These "strings" now represent the approximate slope of the integral curve across the respective rectangles.

Beginning at the origin, in the first rectangle, draw a line parallel to the *string* derived from its average ordinate and terminating at the right edge of the rectangle. From this point, draw the next line parallel to the *string* derived from the average ordinate of the next rectangle and again terminating at its right edge. Continue the process across all of the selected rectangles.

Finally, draw a smooth curve through the intersections of these lines with the edges of the rectangles. This curve will be a close approximation of the integral curve.

The calibration, or scaling of the X-axis, is common to both curves. The calibration of the Y-axis of the integral scale is most easily done by using the same unit distance of the basic curve and recognizing that the value of the same unit distance on the integral scale will be such that:

value of unit on integral scale Y
= (value of unit distance on original Y-axis)
  $\times$ (value of unit distance on X-axis)
  $\times$ (pole distance in unit distances of X)

The total area under the curve, between the limits specified, is then indicated by the maximum ordinate of the integral curve. But since the integral curve represents a continuous summation of incremental areas, the ordinate of the integral curve at *any* point represents the net area under the base curve from the initial limit up to that point. This is extremely valuable in analyzing the relationships of the variables at *any* point on the curve.

144

Figure 10-5   Integral Curve (S vs. T Curve) Drawn on Separate Axes

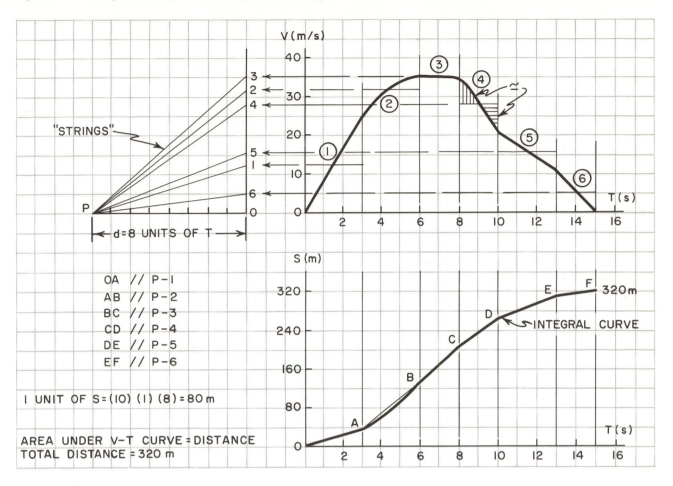

"STRINGS"

d=8 UNITS OF T

OA // P-1
AB // P-2
BC // P-3
CD // P-4
DE // P-5
EF // P-6

1 UNIT OF S = (10) (1) (8) = 80 m

AREA UNDER V-T CURVE = DISTANCE
TOTAL DISTANCE = 320 m

INTEGRAL CURVE

Figure 10-6   Original (F vs. D) Curve and Integral (W vs. D) Curve on Same Axes

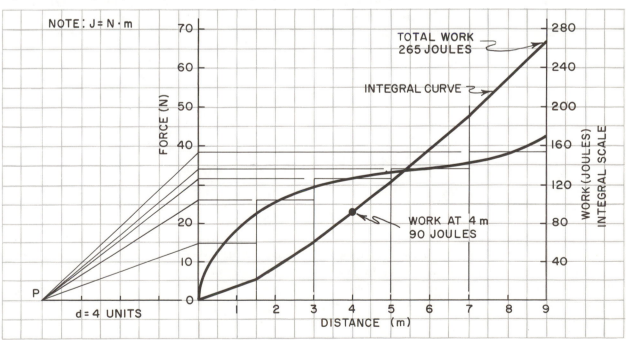

NOTE: J ≡ N·m

TOTAL WORK
265 JOULES

INTEGRAL CURVE

WORK AT 4 m
90 JOULES

d = 4 UNITS

145

10.7 SEMIGRAPHICAL INTEGRATION

The *definite integral of a function* is defined mathematically as the limit of the sum

$$\lim_{n \to \infty} S_n = \lim_{n \to \infty} \sum_{k=1}^{n} f(x_k^*) \, \Delta x_k = \int_{x_0}^{x_n} f(x) \, dx$$

Geometrically $f(x_k^*)$ is some value of $f(x)$ between x_{k-1} and x_k. Thus, in Figure 10-7, $f(x_1^*)$ is some value of y between y_0 and y_1 at $x = x_1^*$ which lies on the curve shown. The product $f(x_1^*)\Delta x_1$ may then be considered as the *area* of a rectangle of base Δx_1 and *average* height $f(x_1^*)$. The sum S_n is thus the sum of the areas of a series of adjacent rectangles or the *entire area* between the curve and the x axis, and $y = x_0$ and x_n.

If the number of rectangles is finite, the limit S_n will approximate the total area to the extent that the heights of the rectangles—$f(x_k^*)$—differ from the exact average height of each rectangle. Thus, if the height of each rectangle is taken at its left-hand edge as $y_{k-1} = f(x_{k-1})$, the area will be that inscribed below the curve or area *abcdefghij*. If, however, the heights are taken at the right-hand edge as $y_k = f(x_k)$, the area will be that circumscribing the curve or area $ab'c'd'e'f'g'h'k'j$. This entire area could also be equated to an average height, y_{av}, of the entire area times the base $x_n - x_0$. If $n \to \infty$, either $f(x_{k-1})$ or $f(x_k)$ may be used and the limit S_n will be the *exact* value of the integral between $x_0 < x < x_n$, assuming that $f(x)$ is continuous over this interval also.

A close approximation for the value of this *definite integral $\int_{x_0}^{x_n} f(x) \, dx$*, as it is called, may be obtained by assuming that $f(x_k^*)$ varies linearly with x. Geometrically, this means that the area is approximated by a series of trapezoids whose overall average height is

$$y_{av} = \frac{1}{2n} \left[y_0 + 2(y_1 + y_2 + \cdots + y_{n-1}) + y_n \right]$$

if all the Δx's are made equal. Then the area A is

$$A = n \, \Delta x \, y_{av} = \int_{x_0}^{x_n} f(x) \, dx$$

This equation is known as the *trapezoidal rule*.

An even closer approximation may be obtained by assuming that every three adjacent ordinates like y_0, y_1, and y_2 are connected by a second-order curve $y = ax^2 + bx + c$. Then if all the Δx's are equal, the overall average height is

Figure 10-7

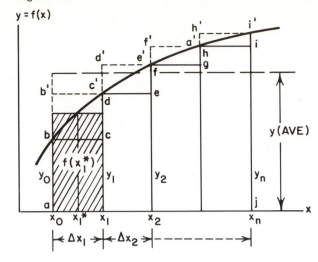

$$y_{av}^* = \frac{1}{3n} (y_0 + 4y_1 + 2y_2 + 4y_3 + 2y_4$$

$$+ \cdots + 2y_{n-2} + 4y_{n-1} + y_n)$$

where n is even, that is, the interval $n\Delta x = x_n - x_0$ must be divided into an *even number* of equal increments. Then the area A is

$$A = n \, \Delta x \, y_{av}^* = \int_{x_0}^{x_n} f(x) \, dx$$

This equation is known as *Simpson's rule*.

It should be noted that both of these rules are based on known values of y_k's at equal intervals of Δx. If the measured data for one variable are not taken at equal intervals as in Figure 10-2, a smooth curve must be plotted and the y_k's measured on this graph at equal intervals of Δx.

This graphical method of evaluating definite integrals may also be used when $y = f(x)$ is a *known function but when the integral is too difficult to evaluate mathematically*. In that case $f(x_k)$ should be evaluated for a series of arbitrarily selected values of x_k that differ by equal increments Δx from adjacent abscissa.

Example. Determine the area under the curve of Figure 10-2 and indicate its proper units. (Note: This area, $\int_0^\epsilon f\sigma \, d\epsilon$, measures the energy of rupture U_r, that is, toughness.) Use Simpson's rule.

Solution. Divide the abscissa ϵ into an *even* number of increments, say 6. Read σ vs. ϵ and record these values as follows:

σ	0	55	90	125	154	172	186
ϵ	0	0.0004	0.0008	0.0012	0.0016	0.0020	0.0024

146

Then

$$\sigma^*_{av} = \frac{1}{3(6)} [0 + 4(55 + 125 + 172)$$

$$+ 2(90 + 154) + 186]$$

$$= 115.67 \text{ MPa}$$

and

$$U_r = n \, \Delta \epsilon \, \sigma^*_{av} = 6(0.0004 \text{ m/m})(115.67 \text{ MN/m}^2)$$

$$= 0.2776 \text{ MN} \cdot \text{m/m}^3$$

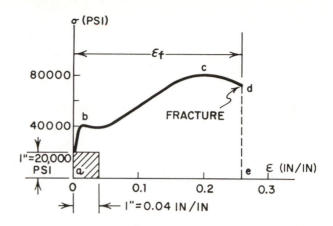

Figure 10–8

10.8 OTHER TECHNIQUES FOR AREA MEASUREMENT

Several techniques are available for measuring areas. The simplest of these involves a *weighing technique*. Here, if the area is cut out of a piece of cardboard or sheet metal and weighed, and a unit square (say, a piece 1 in. by 1 in.) is also cut and weighed, their quotient will give the area in square units. Care must be taken to use cardboard or sheet metal that is uniform in thickness and homogeneous, and the weights must be determined to at least three significant figures.

Another simple method for determining the magnitude of an area is to use a *planimeter*. This is an instrument that measures the area in square inches by tracing a pointer, attached to the instrument, completely around the perimeter of the area in a clockwise direction.

Whether one uses the weighing or planimeter technique, the area must be evaluated in its proper units.

Example. A ductile steel bar is pulled in a tensile testing machine to fracture. The load and accompanying elongation are plotted automatically to give the stress-strain curve shown in Figure 10-8. The scales used are 1 in. of ordinate = 20,000 psi of stress (σ) and 1 in. of abscissa = 0.04 in./in. of strain (ϵ). The area under the curve (*abcdea*) is measured with a planimeter and found to be 15.6 in.^{2} What is the energy of rupture?

Solution. The energy of rupture is defined as $U_r = \int_0^\epsilon f \sigma d\epsilon$, which is the area under the σ-ϵ curve to fracture. This area was measured as 15.6 in.^{2}. Now, for the scales used, one square inch of area is

$$20,000 \text{ lb./in.}^2 \times 0.04 \text{ in./in.} = 800 \text{ in. lb./in.}^3$$

Then

$$U_r = 15.6 \times 800 = 12,480 \text{ in. lb./in.}^3$$

Index

150

Sketch the vertical capital letters and numerals. First sketch lightly over the given copy and then sketch one copy below. Memorize the shapes and proportions.

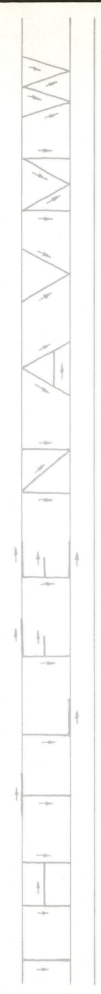

H T I L F E N A V M W

K X Y Z U J O O C G

D P R B S & 1 2 3 4 5

$2\frac{3}{4}$ $4\frac{15}{16}$ $\frac{9.2505}{9.2500}$

6 7 8 9 0

Sketch the lower case letters below. First sketch lightly over the given copy and then sketch copies below. Memorize the shapes and proportions.

i t v w x y z k r j

o c a b d p q g e h n

u m f s   Lower Case Letters

This type of lettering is commonly used on drawings

You may find it more convenient to start your letters like this

SECTION	TABLE
COURSE	DATE

DR. NO.
1-2

DRAWN BY
INSTRUCTOR

Letter the data listed below. Use guidelines for 3 mm letters. Make the letters with firm, uniform strokes of the pencil.

NAME OF YOUR HIGH SCHOOL: YEAR GRADUATED:

LOCATION OF HIGH SCHOOL:

HOME ADDRESS: TELEPHONE:

YOUR PRESENT COLLEGE ADDRESS   TELEPHONE:

Letter the map title balanced on the given center line. Draw thin gray lines with a lettering guide for the millimeter sizes listed. Allow 10 mm vertical spacing between lines.

```
            CONTOUR MAP              (6 mm)
      TALL OAKS GOLF COURSE          (7 mm)
         NEWTOWN, VIRGINIA           (5 mm)
   SCALE 1:1000  SEPTEMBER 15, 1987  (4 mm)
```

SECTION		TABLE	
COURSE		DATE	

DRAWN BY
INSTRUCTOR

DR. NO.
I-3

In the space to the right reproduce freehand the given three views and the isometric view of the block. Identify the points on your drawings with letters corresponding to the letters on the given views. Be particularly careful to maintain parallelism of lines.

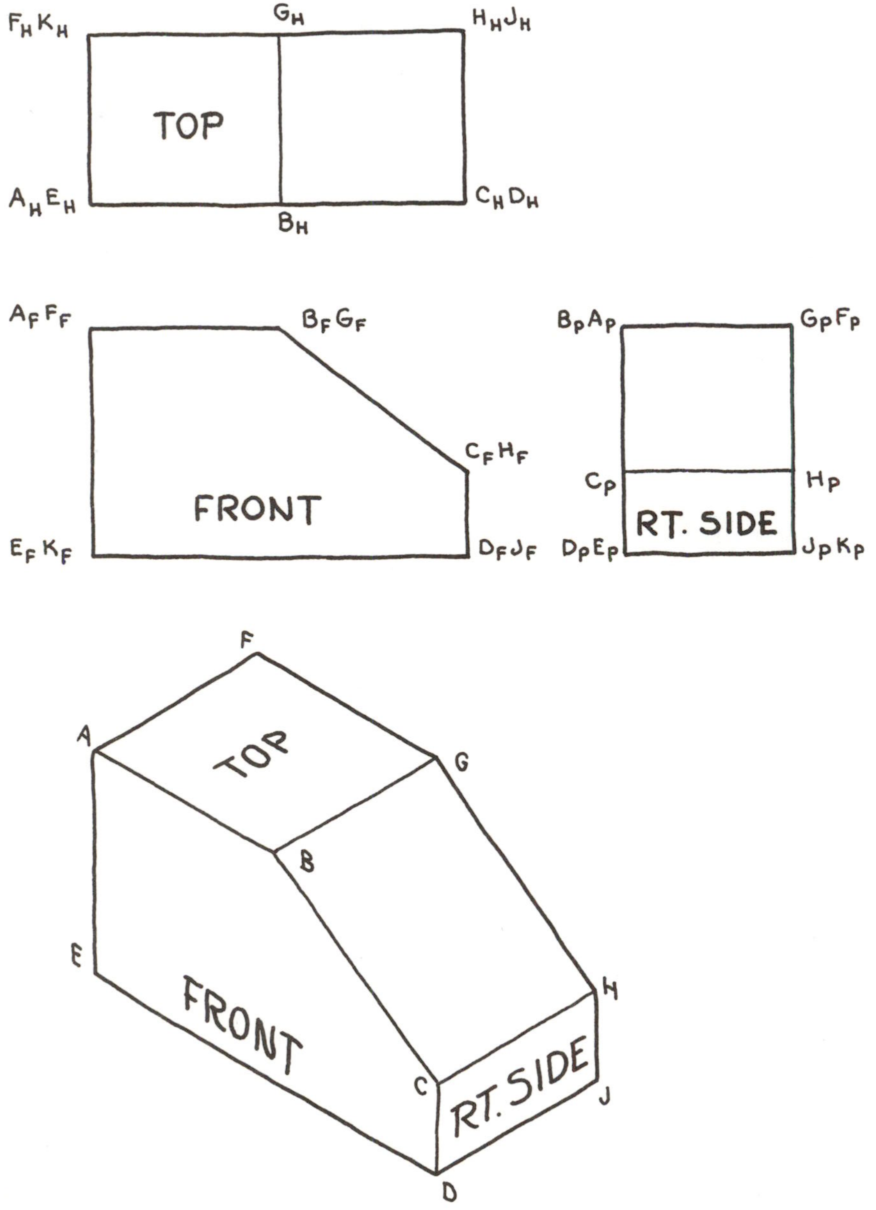

TOP

$F_H K_H$ G_H $H_H J_H$ $C_H D_H$

$A_H E_H$ B_H

FRONT

$A_F F_F$ $B_F G_F$ $C_F H_F$ $D_F J_F$

$E_F K_F$

RT. SIDE

$B_P A_P$ $G_P F_P$ H_P $J_P K_P$

C_P $D_P E_P$

TOP

FRONT

RT. SIDE

F   G   H

A   B   C   D   J   E

SECTION   TABLE   DRAWN BY
COURSE   DATE   INSTRUCTOR

DR. NO.
1-4

Given an isometric drawing of a block.

In the space to the right, at approximately the same size, sketch freehand the front, top, and right side (profile) views of the block. Label height, width, depth, and points of the block.

Note: An isometric drawing is one method of representing an object pictorially.

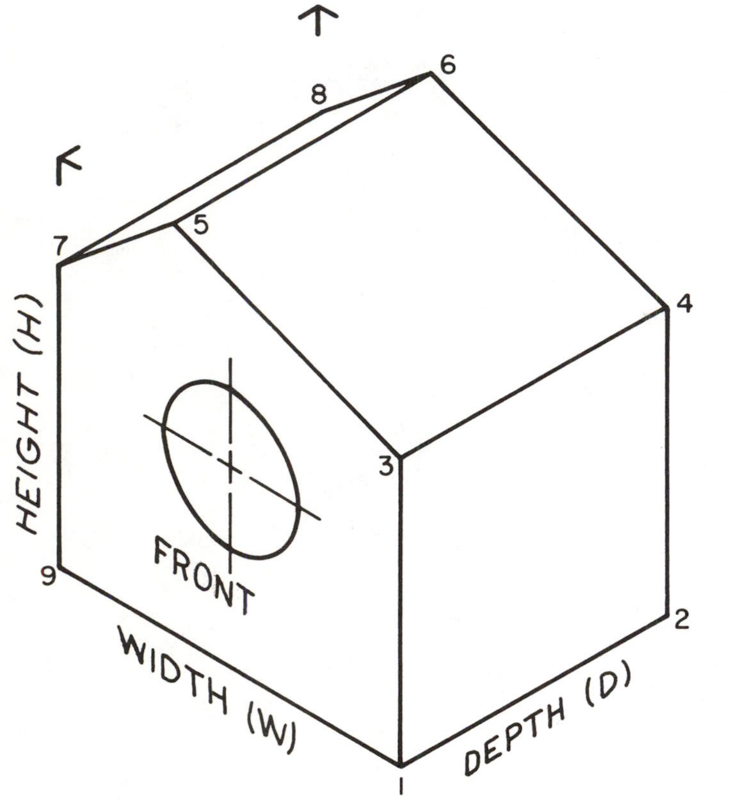

HEIGHT (H)

WIDTH (W)

DEPTH (D)

FRONT

| SECTION | TABLE | DRAWN BY | DR. NO. |
| COURSE | DATE | INSTRUCTOR | 1-5 |

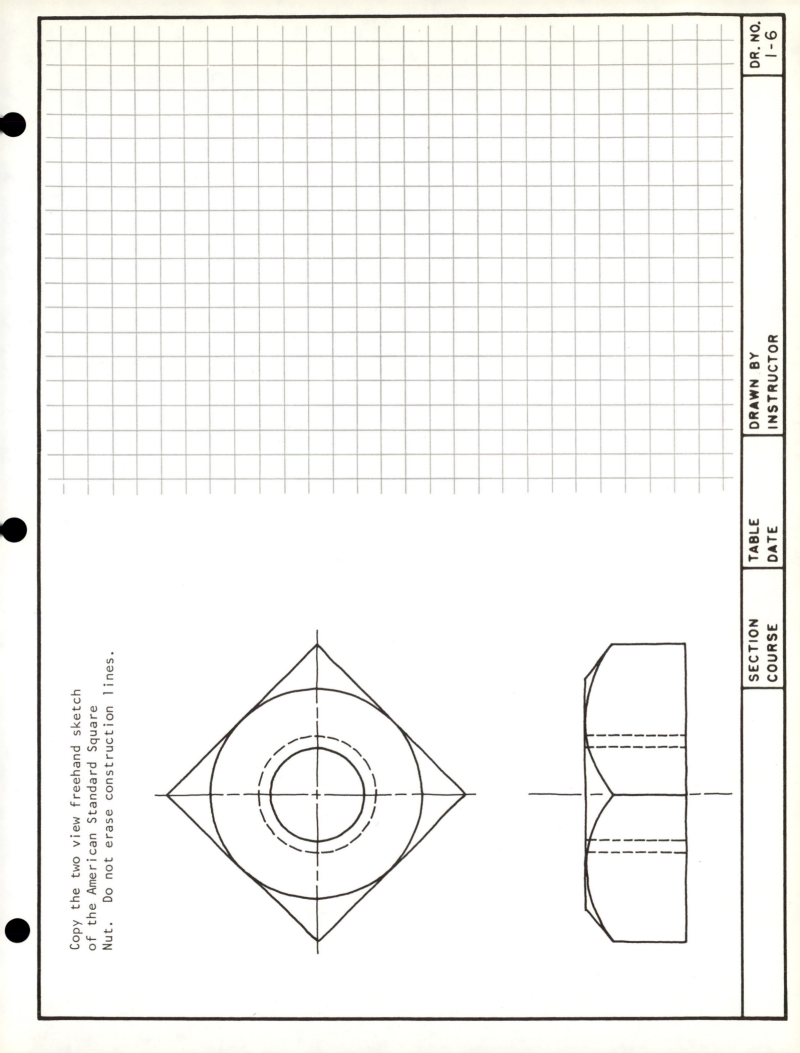

Copy the two view freehand sketch
of the American Standard Square
Nut. Do not erase construction lines.

DRAWN BY
INSTRUCTOR

TABLE
DATE

SECTION
COURSE

Sketch freehand isometric pictorial views of (1) a cube, (2) a triangular prism, (3) a right pyramid and (4) a cylinder. Use the same size box construction for all objects and retain all construction lines.

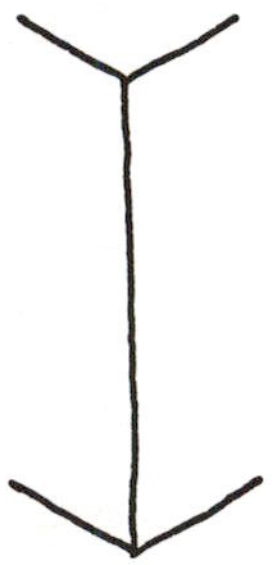

2

4

1

3

| SECTION | TABLE | DRAWN BY |
| COURSE | DATE | INSTRUCTOR |

DR. NO.
1-7

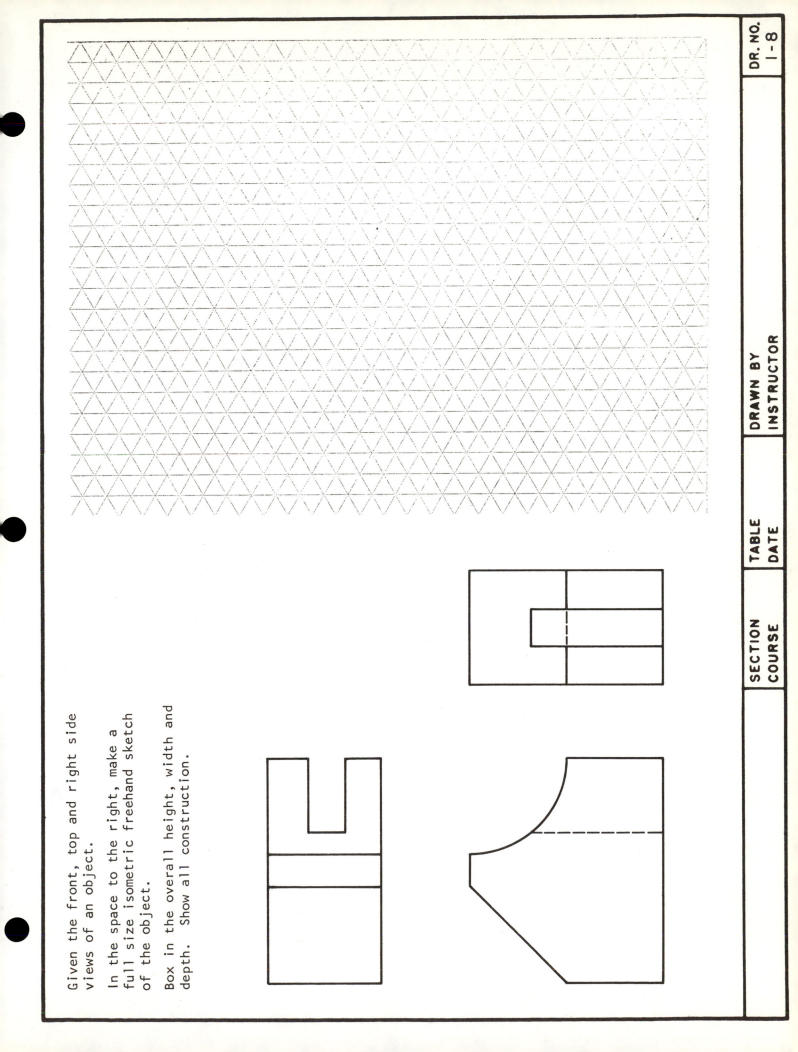

Given the front, top and right side
views of an object.

In the space to the right, make a
full size isometric freehand sketch
of the object.

Box in the overall height, width and
depth. Show all construction.

DRAWN BY
INSTRUCTOR

TABLE
DATE

SECTION
COURSE

In the space to the right, sketch freehand the front, top and right side views of the object shown below.

Increase the size 50% by estimation.

Show all construction.

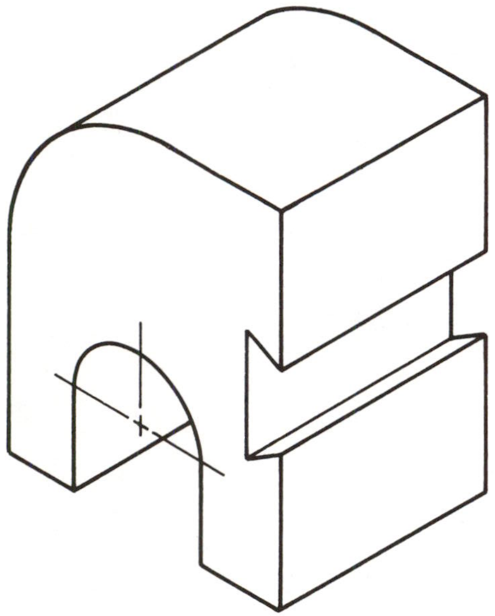

	SECTION	TABLE	DRAWN BY	DR. NO.
	COURSE	DATE	INSTRUCTOR	1 - 9

In the space to the right, make an isometric freehand sketch of the object shown below.

Double the size by estimation.

Show all construction.

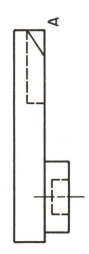

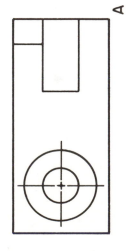

| SECTION | TABLE | DRAWN BY | DR. NO. |
| COURSE | DATE | INSTRUCTOR | 1-10 |

Sketch freehand oblique pictorial views of (1) a cube, (2) a triangular prism, (3) a right pyramid and (4) a cylinder. Use the same size box construction for all objects and retain all construction lines.

1

2

3

4

| SECTION | TABLE | DRAWN BY | DR. NO. |
| COURSE | DATE | INSTRUCTOR | 1-11 |

In the space to the right, make an oblique freehand sketch of the object shown below. Increase the size 100% by estimation.

Show all construction.

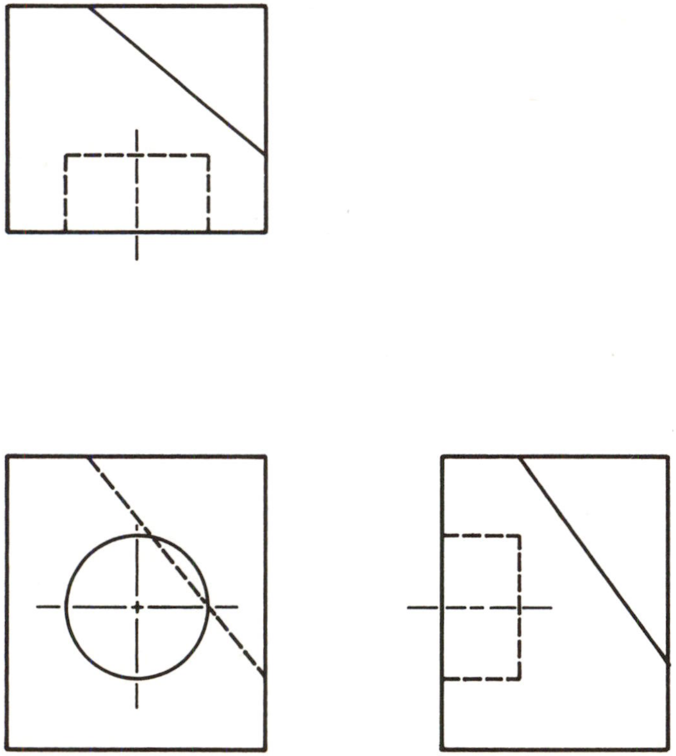

	SECTION COURSE	TABLE DATE	DRAWN BY INSTRUCTOR	DR. NO. 1-12

On the isometric drawing below, box in the given object. The right rear corner of the box is indicated.

In the space to the right, make a full size oblique freehand sketch of the object.

Show all construction.

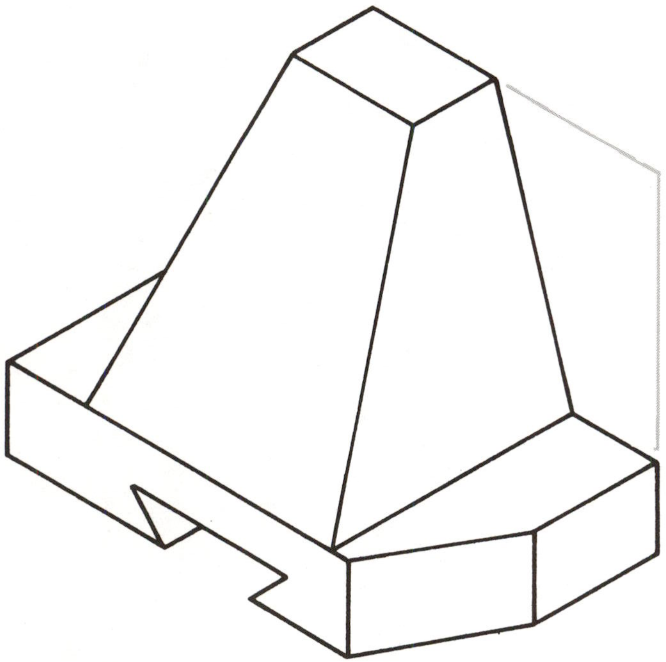

SECTION
COURSE

TABLE
DATE

DRAWN BY
INSTRUCTOR

DR. NO.
1-13

In the space to the right, make a full
size cabinet freehand sketch of the
object shown below.

Start at the point "A" and use the
given axes. Show all construction.

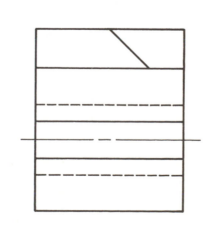

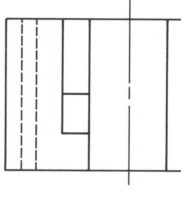

DR. NO.
I-14

SECTION   TABLE   DRAWN BY
COURSE   DATE   INSTRUCTOR

Make a full size freehand pictorial sketch of the object whose front, top, and right side views are shown.

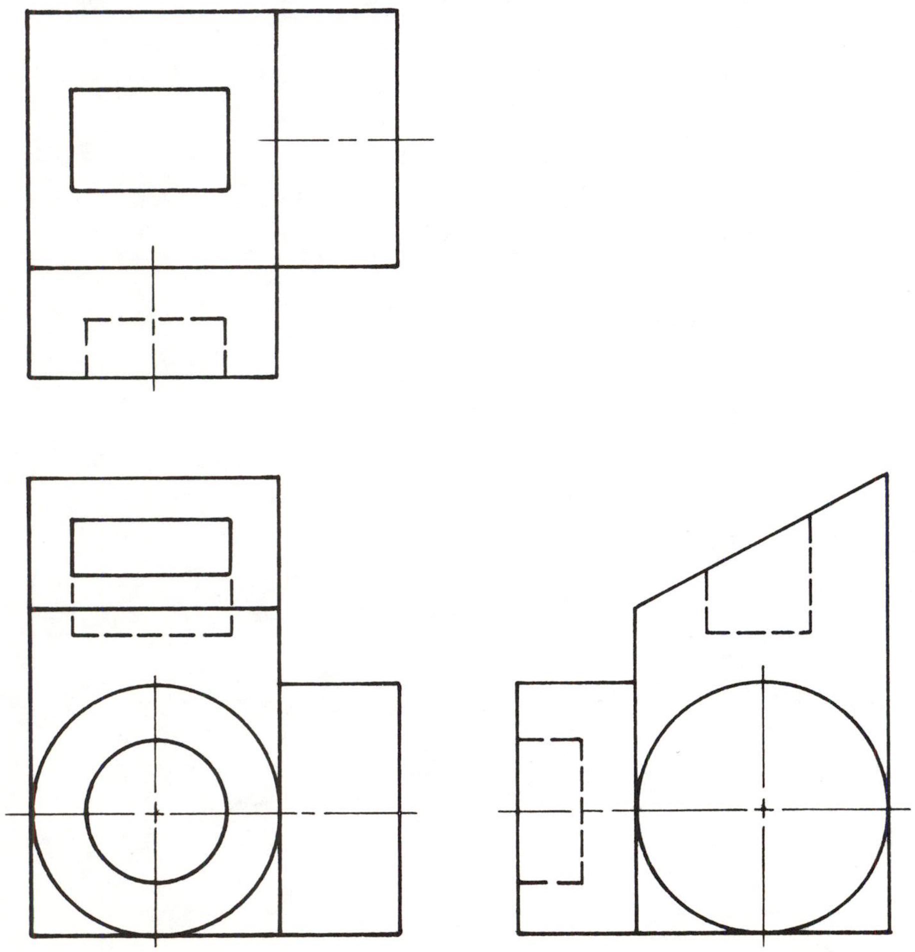

| SECTION | TABLE | DRAWN BY |
| COURSE | DATE | INSTRUCTOR |

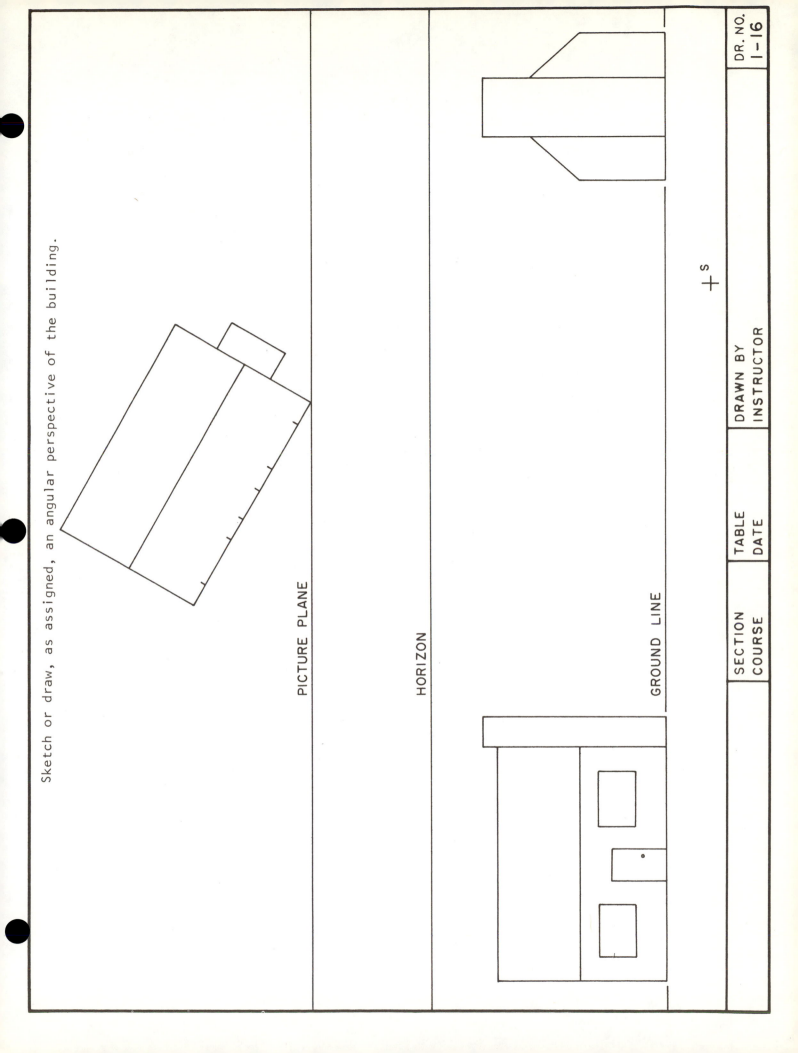

Sketch or draw, as assigned, an angular perspective of the building.

PICTURE PLANE

HORIZON

GROUND LINE

SECTION

COURSE

TABLE

DATE

DRAWN BY

INSTRUCTOR

DR. NO.
1-16

Sketch or draw, as assigned, a parallel perspective and an angular perspective of the monument.

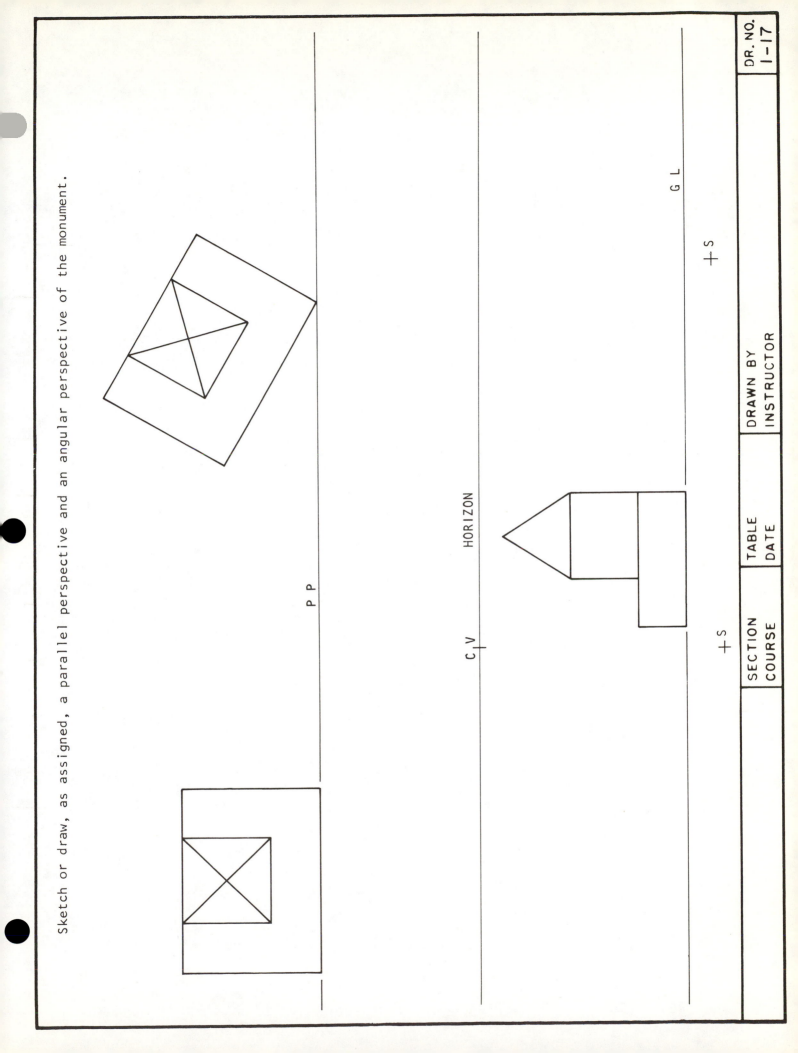

P P

HORIZON

C V

G L

+ S

+ S

SECTION   TABLE   DRAWN BY
COURSE   DATE   INSTRUCTOR

DR. NO.
1-17

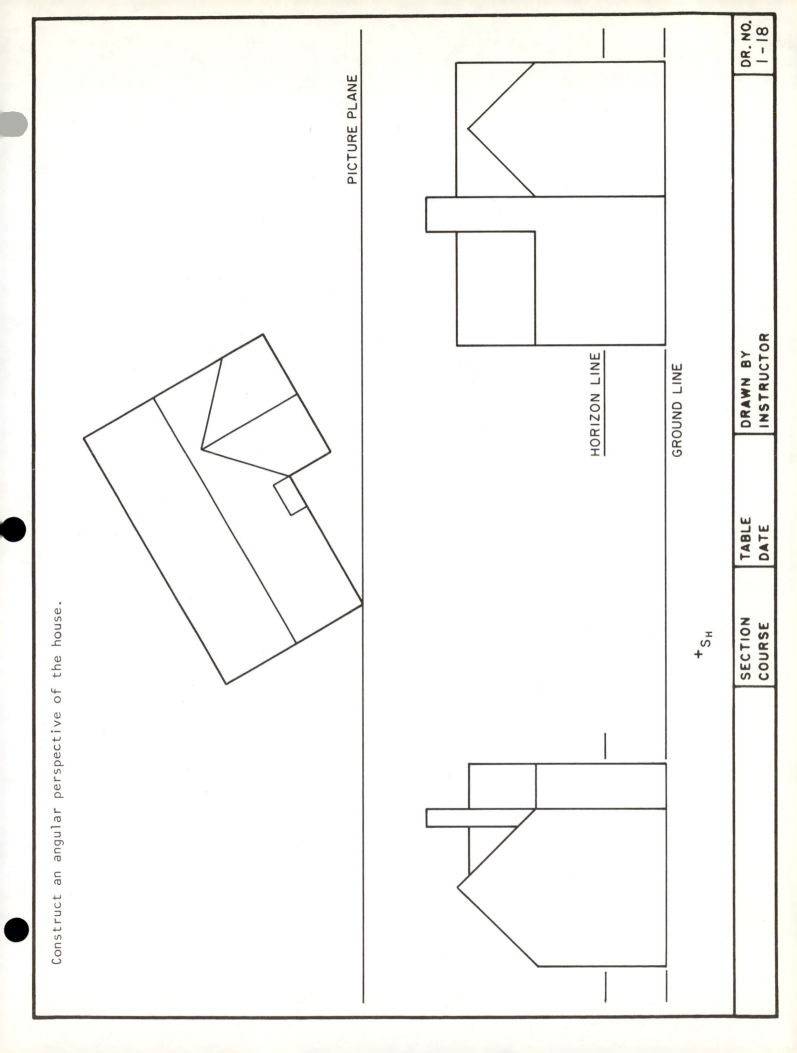

Construct an angular perspective of the house.

PICTURE PLANE

HORIZON LINE

GROUND LINE

$+S_H$

| SECTION | TABLE | DRAWN BY | DR. NO. |
| COURSE | DATE | INSTRUCTOR | I-18 |

In spaces 1, 2, and 3 determine the length of each line with the specified scale and letter the distance or value above the line. Letter the scale ratio where indicated. Read each scale to the nearest division. Measure inside the vertical marks.

Metric Scale

Scale: 1/1
Ratio =

Scale: 1/50
Ratio =

Scale: 1/100
Ratio =

Scale: 1/0.8
Ratio =

Engineer's Scale

Scale:
1" = 600'
Ratio =

Scale:
1" = 30 miles
Ratio =

Scale:
1" = 100 mph

Scale:
1" = 20 lbs

1

2

3

4

Architect's Scale

Scale:
1/4" = 1'-0
Ratio =

Scale:
3/4" = 1'-0
Ratio =

Scale:
3/32" = 1'-0
Ratio =

Scale:
Full Size
Ratio =

Extend the given lines to the specified lengths using the indicated scales.

2.86m

Scale:
1:50

2'-9 1/2"

Scale:
1 1/2" = 1'-0

107 gal

Scale:
1" = 50 gal

18.4m

Scale:
1:200

SECTION	TABLE	DRAWN BY
COURSE	DATE	INSTRUCTOR

DR. NO.
2-1

Using instruments, perform the specified geometric constructions in the five areas below. Show all construction.

1. Construct the perpendicular bisector of line AB. Construct a 15mm radius arc tangent to XY and YZ.

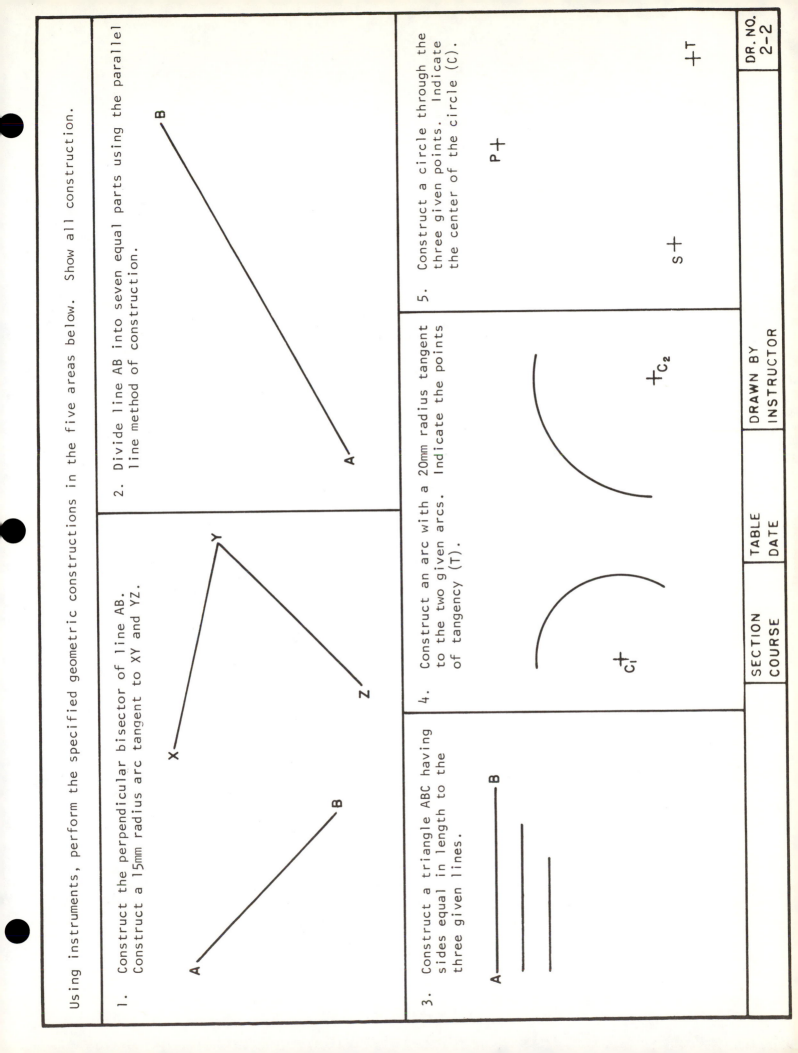

A

B

X

Y

Z

2. Divide line AB into seven equal parts using the parallel line method of construction.

A

B

3. Construct a triangle ABC having sides equal in length to the three given lines.

A ———— B

4. Construct an arc with a 20mm radius tangent to the two given arcs. Indicate the points of tangency (T).

+ C_1

+ C_2

5. Construct a circle through the three given points. Indicate the center of the circle (C).

P +

S +

+ T

SECTION	TABLE	DRAWN BY
COURSE	DATE	INSTRUCTOR

DR. NO.
2-2

A sketch of the center line layout of a proposed race track is shown below. Make an instrument drawing of the layout to a scale of 1/10,000. Show all construction for locating centers and tangent points of the circular curves. Determine the length using 100m chord distances on the curves and state the correction necessary for a 4,000m track. Start the drawing at center point C.

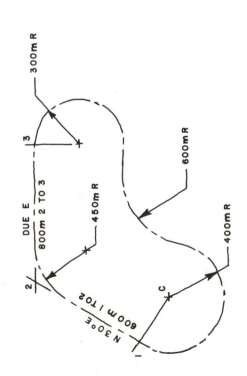

C +

Correction necessary + or − _____ m

N

| SECTION | TABLE | DRAWN BY | DR. NO. |
| COURSE | DATE | INSTRUCTOR | 2-3 |

A country club decided to build a "kidney shaped" swimming pool for its diversified sports recreation area. The architect provides you with a sketch showing the design parameters. Draw the outline of the pool to the scale of 1:200 so that approval may be obtained regarding the shape of the pool before further design layout is pursued. Dimensions are in meters. Show all construction for locating centers and tangent points.

C +

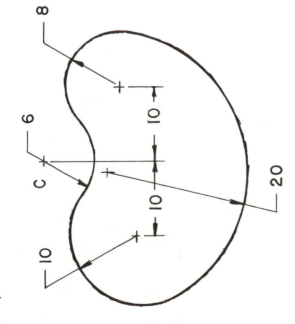

SECTION	TABLE	DRAWN BY
COURSE	DATE	INSTRUCTOR

A block is shown in an isometric drawing.
A 50mm diameter through hole is to be
centered on the stepped face. The radius of
the two incomplete corners is 15mm.

Construct the ellipses necessary to represent
the hole and the missing corners using the
approximate four-center method. Complete the
drawing of the block.

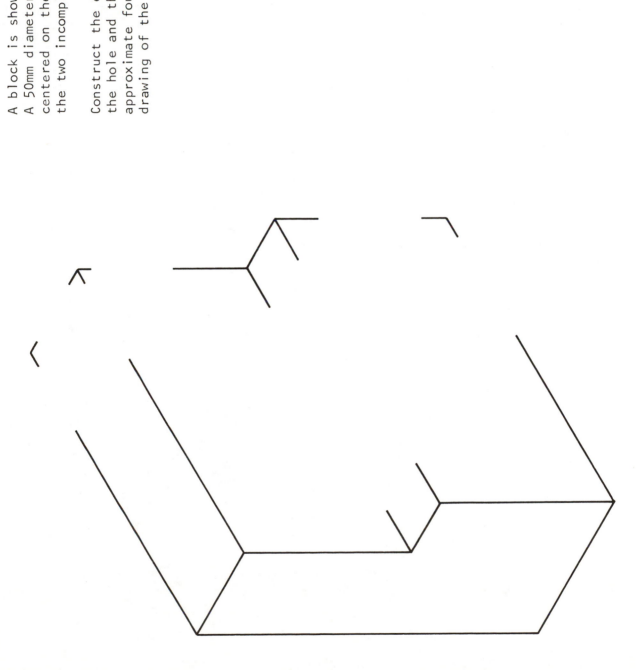

SECTION	TABLE	DRAWN BY	DR. NO.
COURSE	DATE	INSTRUCTOR	2-5

Given an incomplete cavalier drawing of a 70mm solid cube.

Using instruments, complete the given drawing.

1. Construct a hole 40mm in diameter and 15mm deep centered in the front face of the cube. Centered at the bottom of this hole is another hole 30mm in diameter and 15mm deep.

2. Construct a cylinder 50mm in diameter and 25mm high centered on the top face of the cube.

3. Construct a hole 40mm square and 20mm deep centered on the left face of the cube.

4. Show all construction.

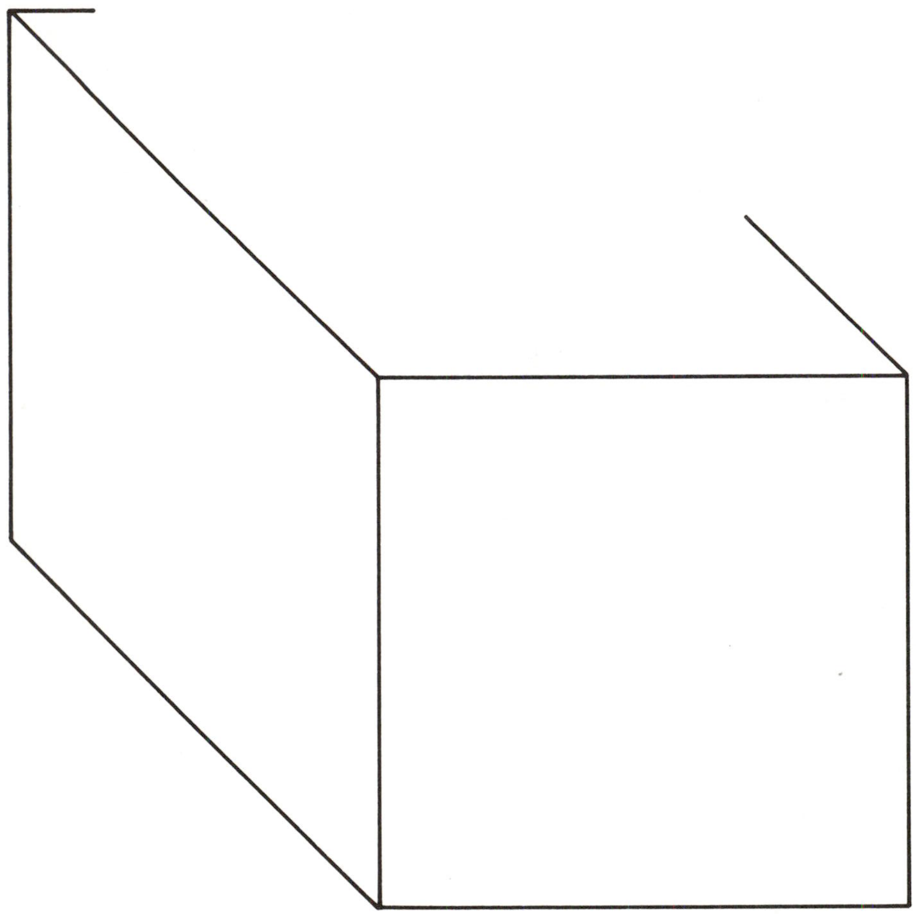

SECTION	TABLE	DRAWN BY	DR. NO.
COURSE	DATE	INSTRUCTOR	2-6

Given an incomplete cavalier oblique drawing
of a 60mm solid cube.

Using instruments, complete the given drawing.

1. Construct a cylinder 40mm in diameter and
    30mm long centered on the front face of
    the cube. This cylinder has a hole 20mm
    square and 15mm deep centered on its axis.

2. Construct a 25mm cube centered on the top
    face of the cube.

3. Construct a hole 30mm square and 15mm deep
    centered on the left face of the cube.

4. Show all construction.

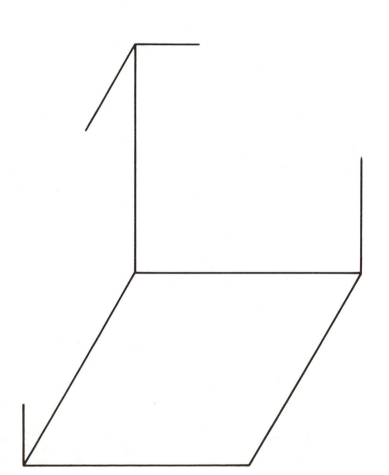

DRAWN BY
INSTRUCTOR

TABLE
DATE

SECTION
COURSE

Given a pictorial view and three orthographic views of a pyramid. Draw the left profile view and letter the projections of the points on all views. Measure the width, depth, and height of the pyramid in millimeters and fill in the dimensions on the drawing. Complete the chart at the right by measuring the numerical values of the coordinates of all points. Scale: 1:2 (half size).

	A	B	C	D	E
BELOW H					
BEHIND F					
LEFT OF P_R					
RIGHT OF P_L					

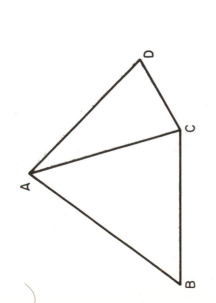

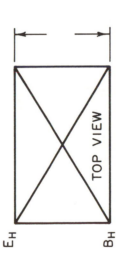

TOP VIEW

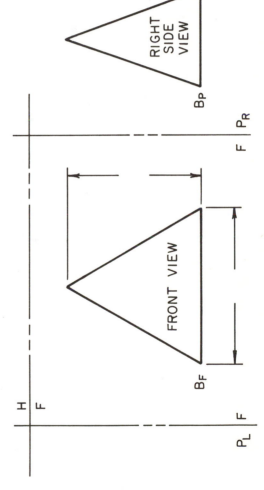

FRONT VIEW

RIGHT SIDE VIEW

| SECTION | TABLE | DRAWN BY | DR. NO. |
| COURSE | DATE | INSTRUCTOR | 3-1 |

Locate A_F and B_P. Measure and record the coordinates of A and B and the distance from A to B. Scale: 1:80.

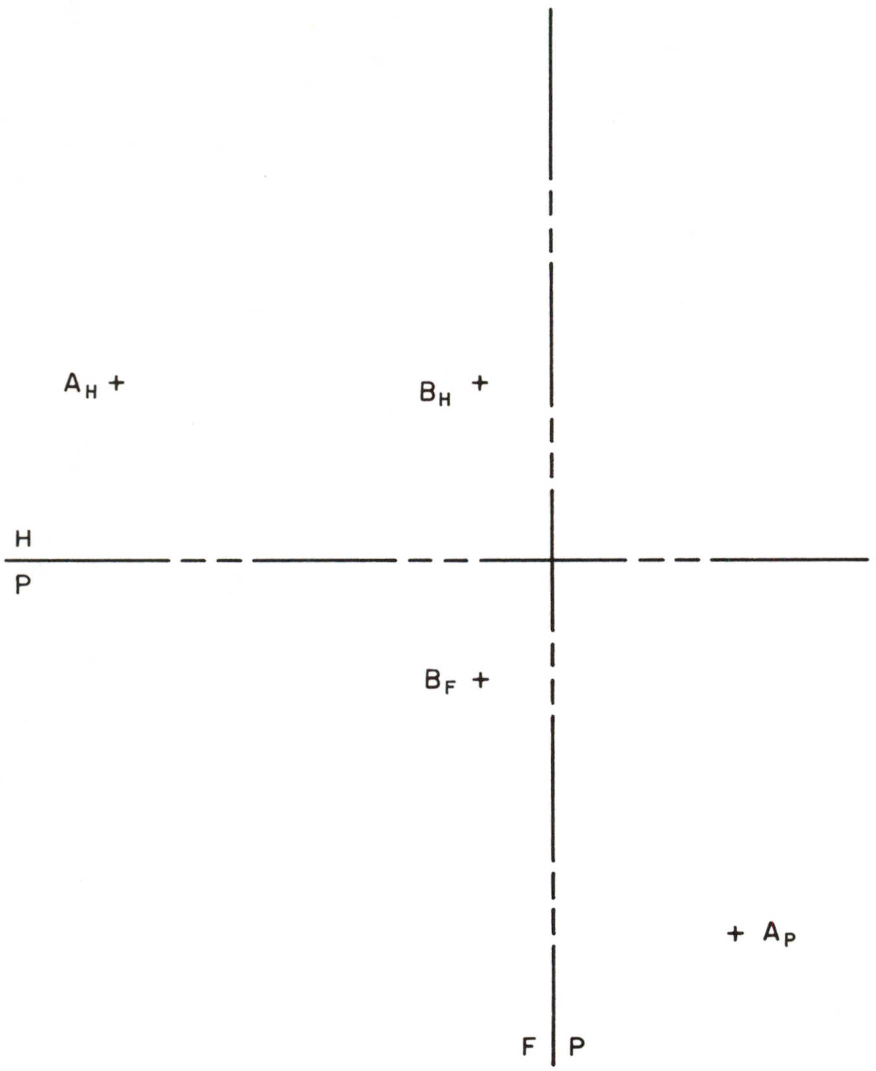

Coordinates	X	Y	Z
Distance From:	P	F	H
Point A			
Point B			

Distance from A to B =

$A_H +$

$B_H +$

$B_F +$

$+ A_P$

H
—
P

F | P

F
P

SECTION	TABLE	DRAWN BY
COURSE	DATE	INSTRUCTOR

The foot of a flag pole is located at point A. The top of the pole, point B, is 16 meters above A. Locate point B and represent the pole with a line. Draw the auxiliary views 1, 2, 3, 4 and 5. List the views that show the true length of the flag pole. Scale: 1:500.

Views which show the true length of the flag pole:

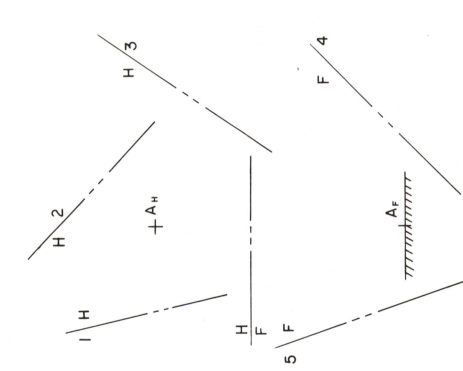

SECTION	TABLE	DRAWN BY
COURSE	DATE	INSTRUCTOR

DR. NO.
3-3

Given two projections of each of three points A, B and C. Locate the five omitted projections of each point.

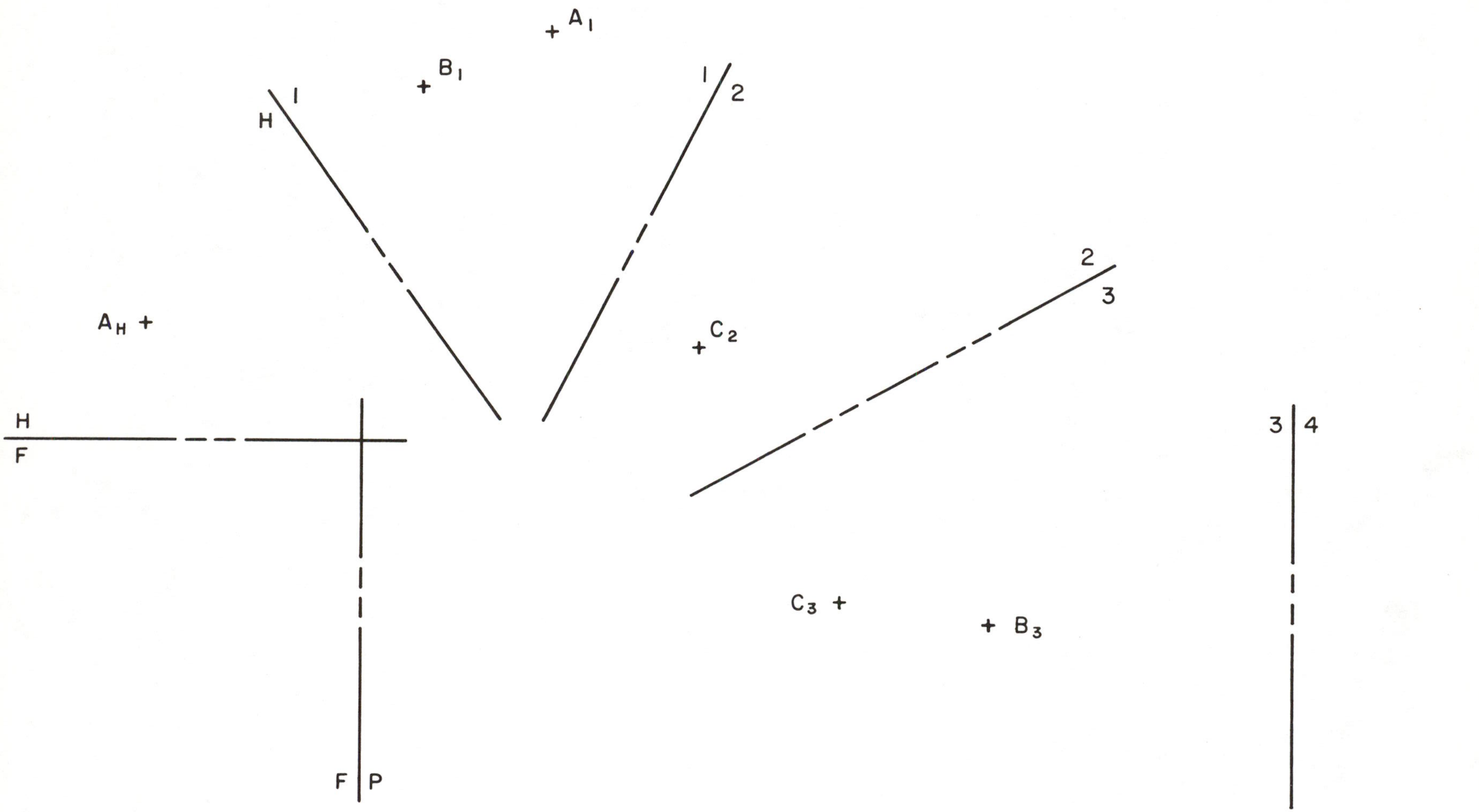

SECTION	TABLE	DRAWN BY
COURSE	DATE	INSTRUCTOR

DR. NO.
3-4

1. Point R is 18 mm behind the frontal plane. Locate R in the top and side views. Express the X (width), Y (depth), and Z (height) coordinates of R (_____ mm, _____ mm, _____ mm).

2. Point S is 30 mm to the right of R, 15 mm above R and 12 mm behind R. Locate S in all views. Express the X, Y, and Z coordinates of S (_____ mm, _____ mm, _____ mm).

3. How far is point S from Point R? Answer = _____ . Indicate this distance on your drawing with the letters TL.

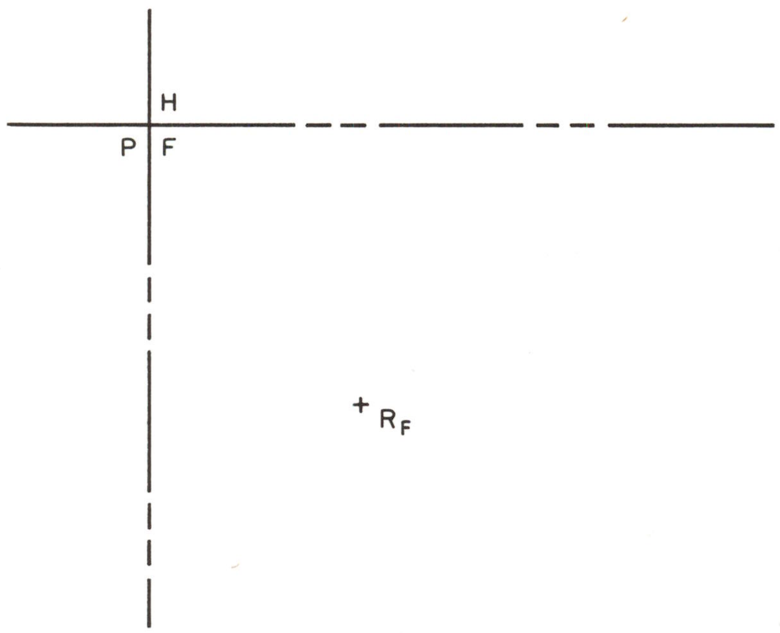

H

P F

+ R_F

SECTION | TABLE | DRAWN BY
COURSE | DATE | INSTRUCTOR

DR. NO.
3-5

Find and label the true length (TL) and the point view (PV) of each of the given lines. Classify each line by checking the proper column in the table. Measure and record the true length of each line. Scale: Full size

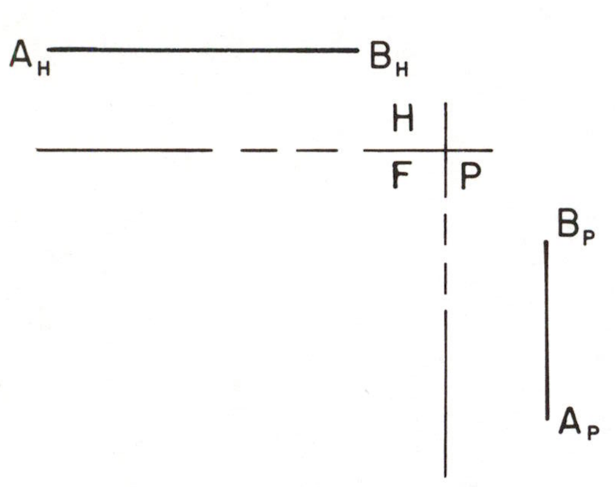

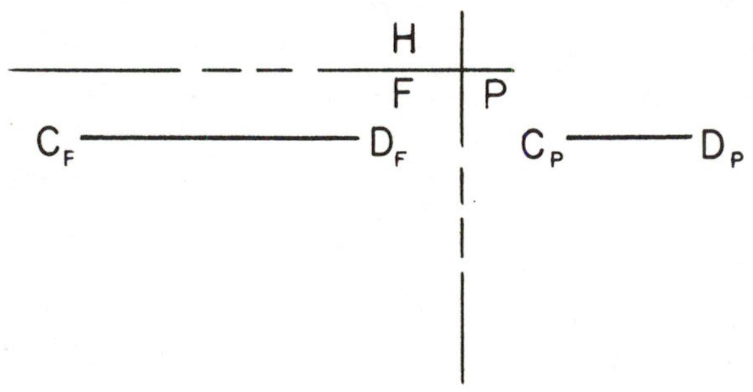

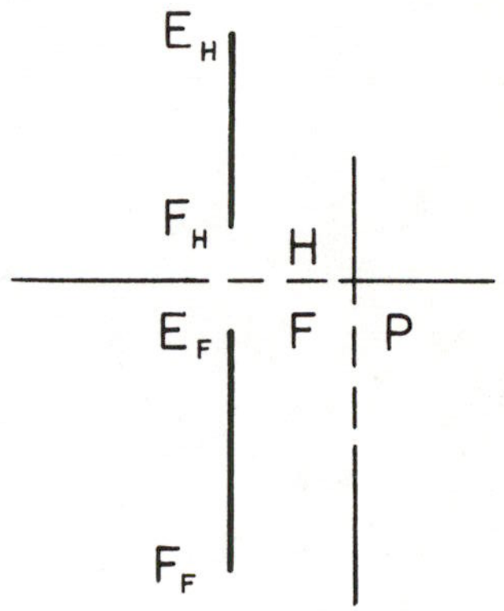

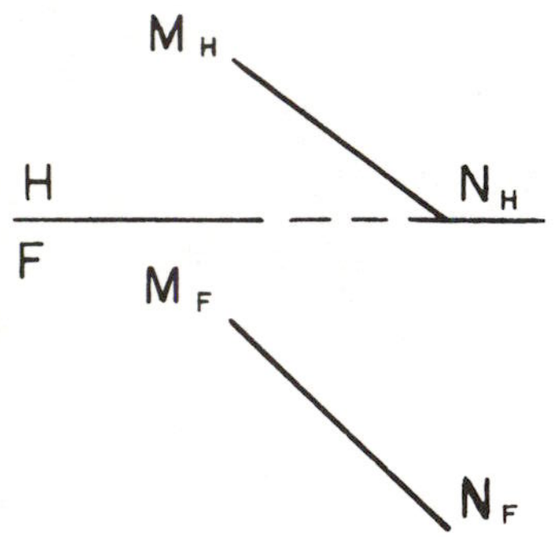

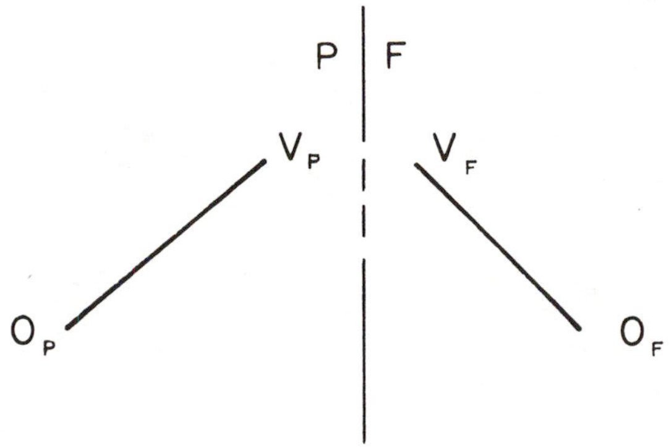

LINE	FRONTAL	HORIZONTAL	PROFILE	OBLIQUE	TRUE LENGTH
AB					
CD					
EF					
MN					
VO					

DR. NO.
3-6

SECTION
COURSE

TABLE
DATE

DRAWN BY
INSTRUCTOR

Locate the missing views of the points and lines which lie in the given planes. Do not use additional views.

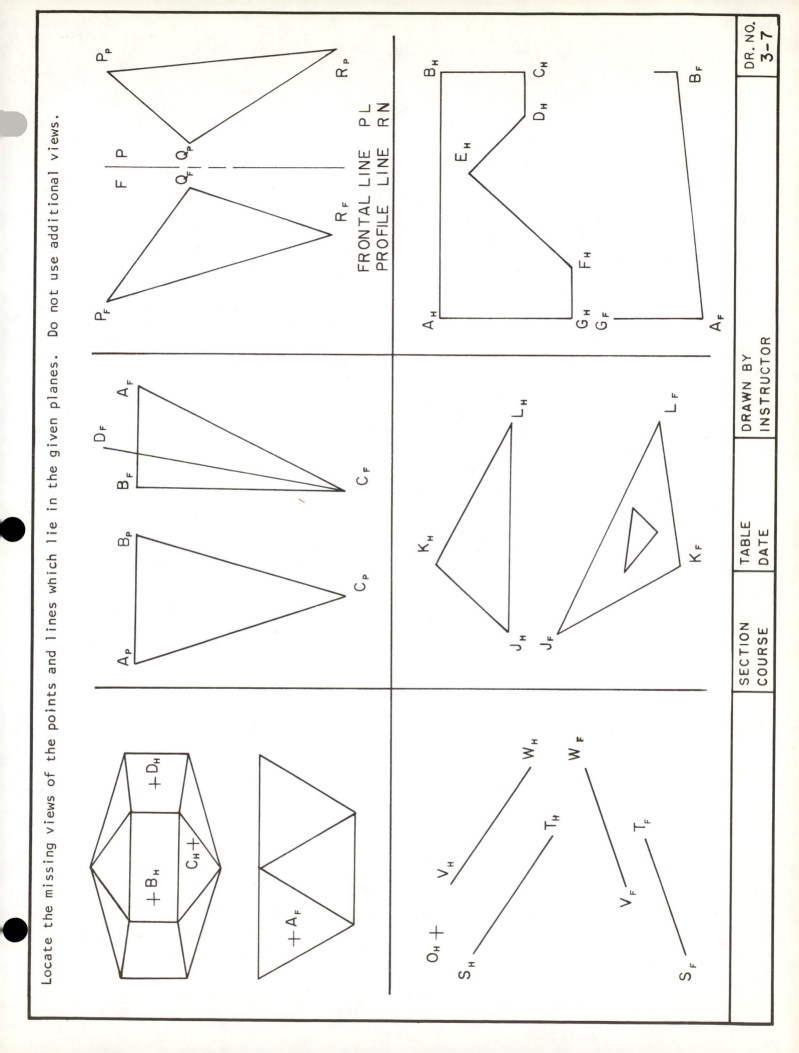

FRONTAL LINE PL
PROFILE LINE RN

DR. NO.
3-7

SECTION   TABLE   DRAWN BY
COURSE   DATE   INSTRUCTOR

1. Find and label the true size of the angle ACB.

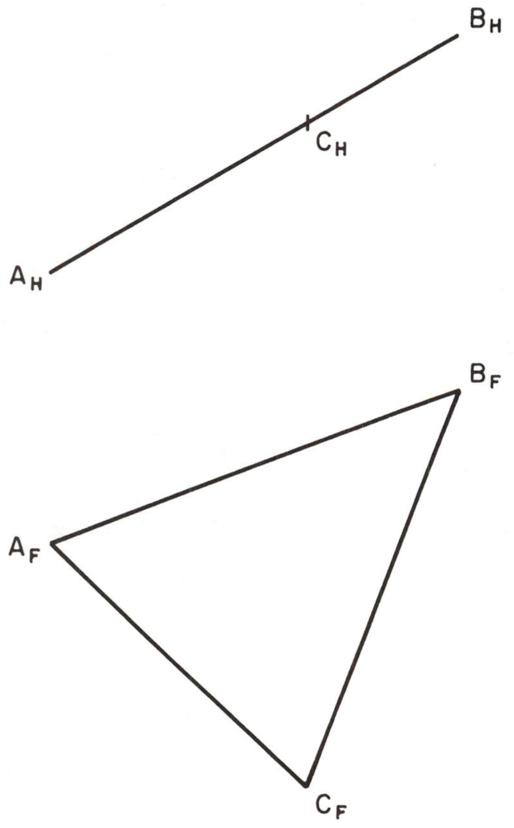

2. Find and label the angle, θ_H, that the plane DEF makes with the horizontal plane. Find and label the true shape view of the plane DEF.

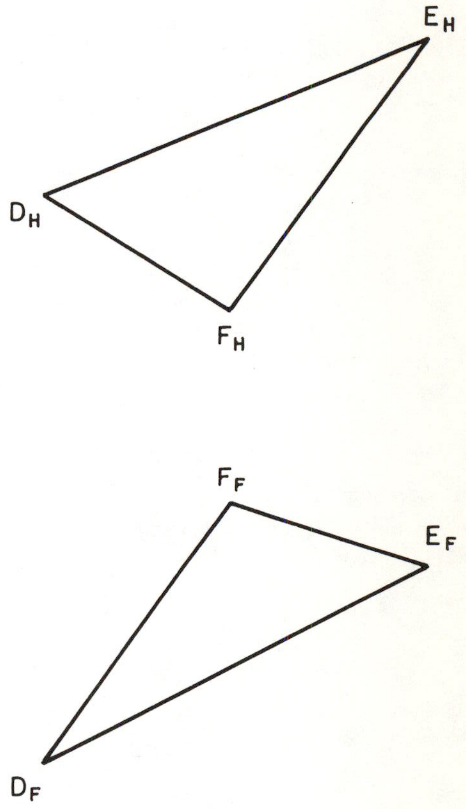

θ_H = _____ °

Angle ABC = _____ °

| SECTION | TABLE | DRAWN BY |
| COURSE | DATE | INSTRUCTOR |

| DR. NO. |
| 3-8 |

Draw the front, top and right side views of the objects illustrated. The front is identified by the letter F. Estimate full size dimensions.

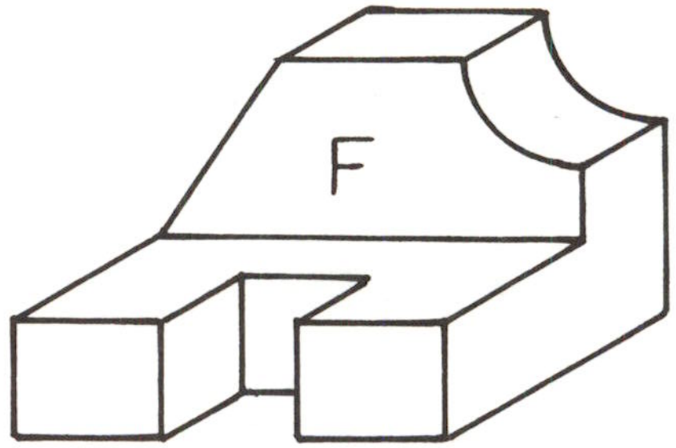

1

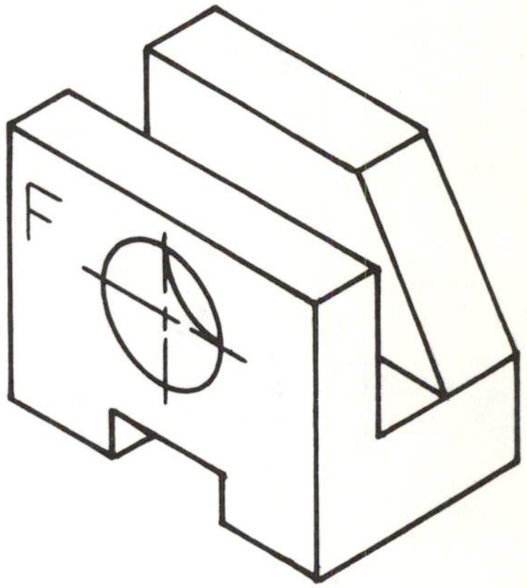

2

3

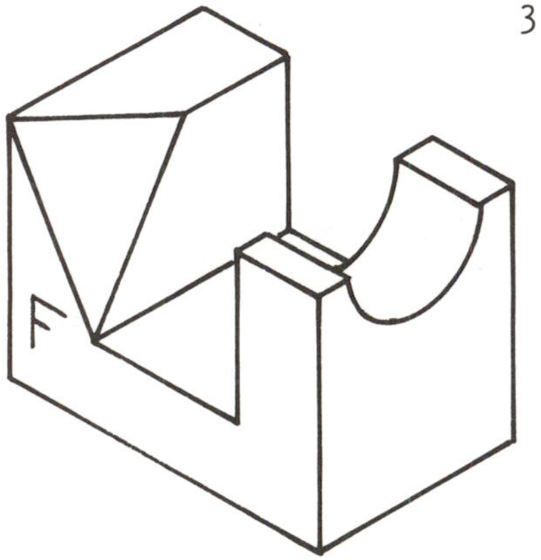

4

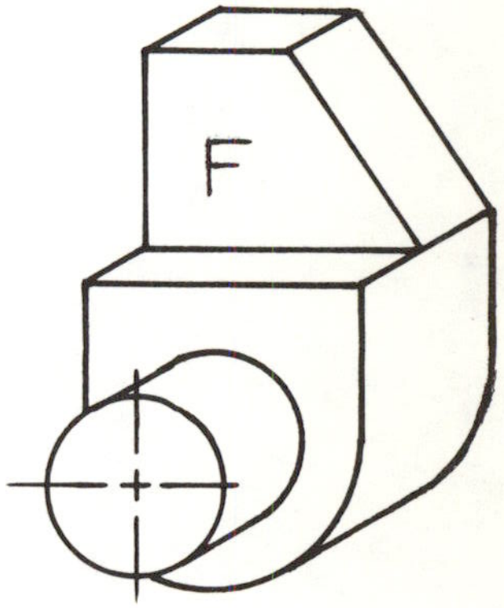

DR. NO.
3-9

DRAWN BY
INSTRUCTOR

SECTION   TABLE
COURSE   DATE

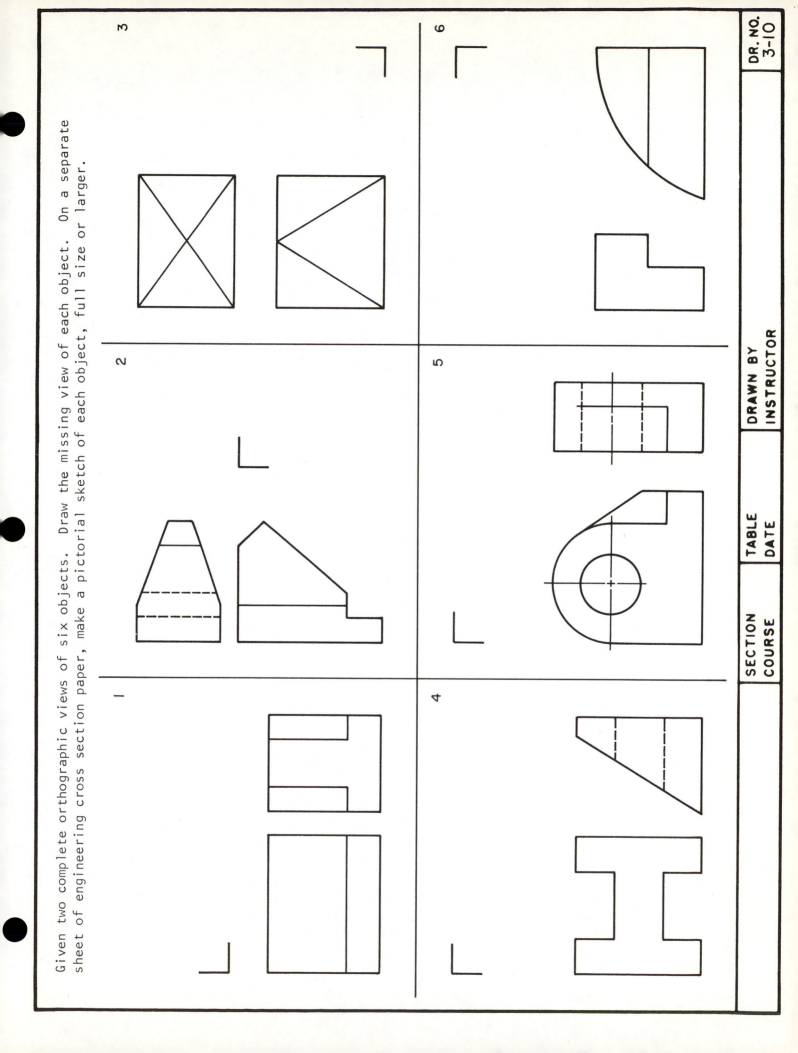

Given two complete orthographic views of six objects. Draw the missing view of each object. On a separate sheet of engineering cross section paper, make a pictorial sketch of each object, full size or larger.

SECTION
COURSE

TABLE
DATE

DRAWN BY
INSTRUCTOR

DR. NO.
3-10

Given two complete orthographic views of six objects. Draw the missing view of each object. On a separate sheet of engineering cross section paper, make a pictorial sketch of each object, full size or larger.

1

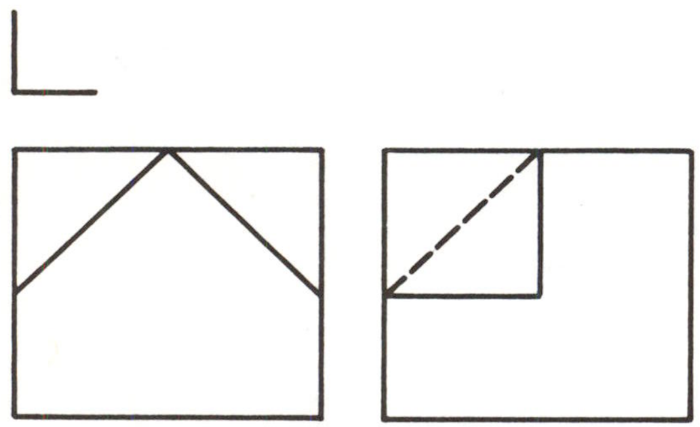

2

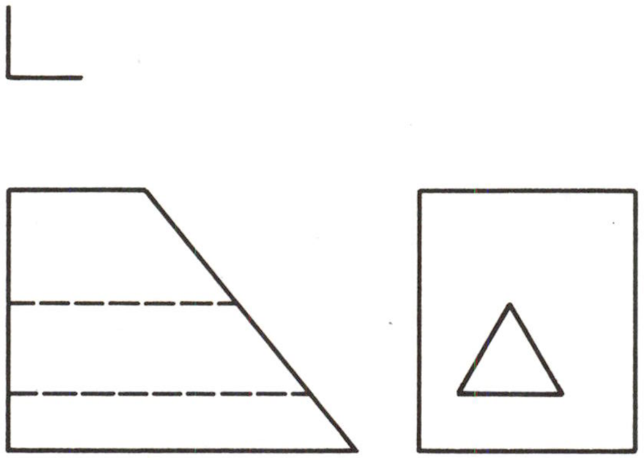

3

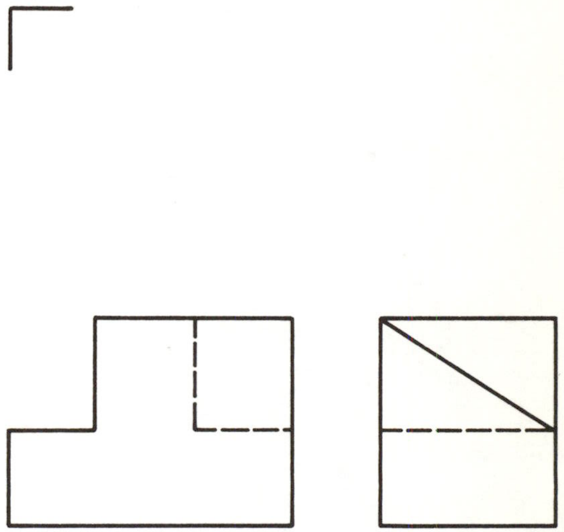

4

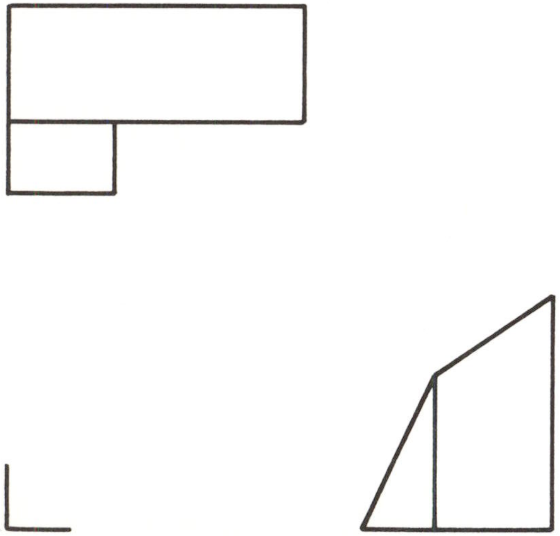

5

6

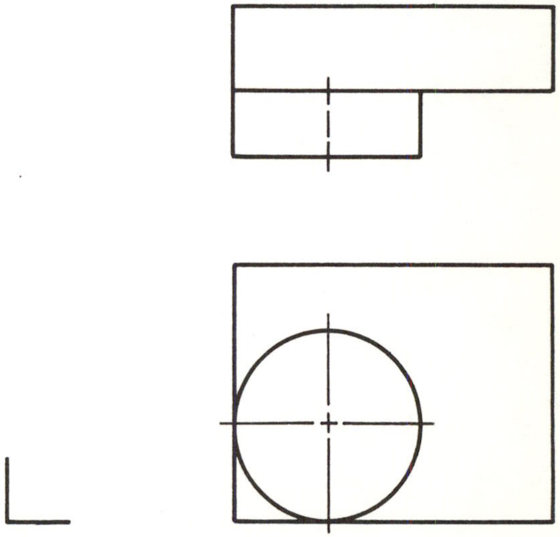

SECTION   TABLE   DRAWN BY
COURSE   DATE   INSTRUCTOR

DR. NO.
3-11

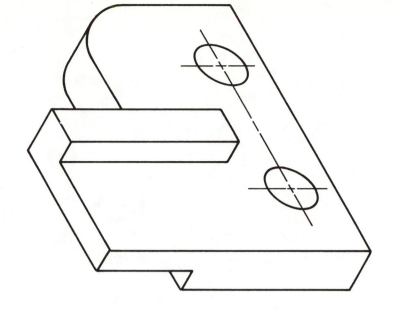

Given the pictorial view of a guide
bracket. Draw the views for a
complete shape description. Transfer
dimensions full size from the pictorial.

| SECTION | TABLE | DRAWN BY |
| COURSE | DATE | INSTRUCTOR |

DR. NO.
3-12

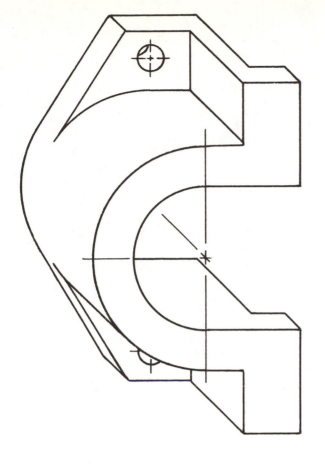

Given a pictorial view of a Pinion Housing. Draw the views for a complete shape description. Transfer dimensions full size from the pictorial.

SECTION
COURSE

TABLE
DATE

DRAWN BY
INSTRUCTOR

DR. NO.
3-13

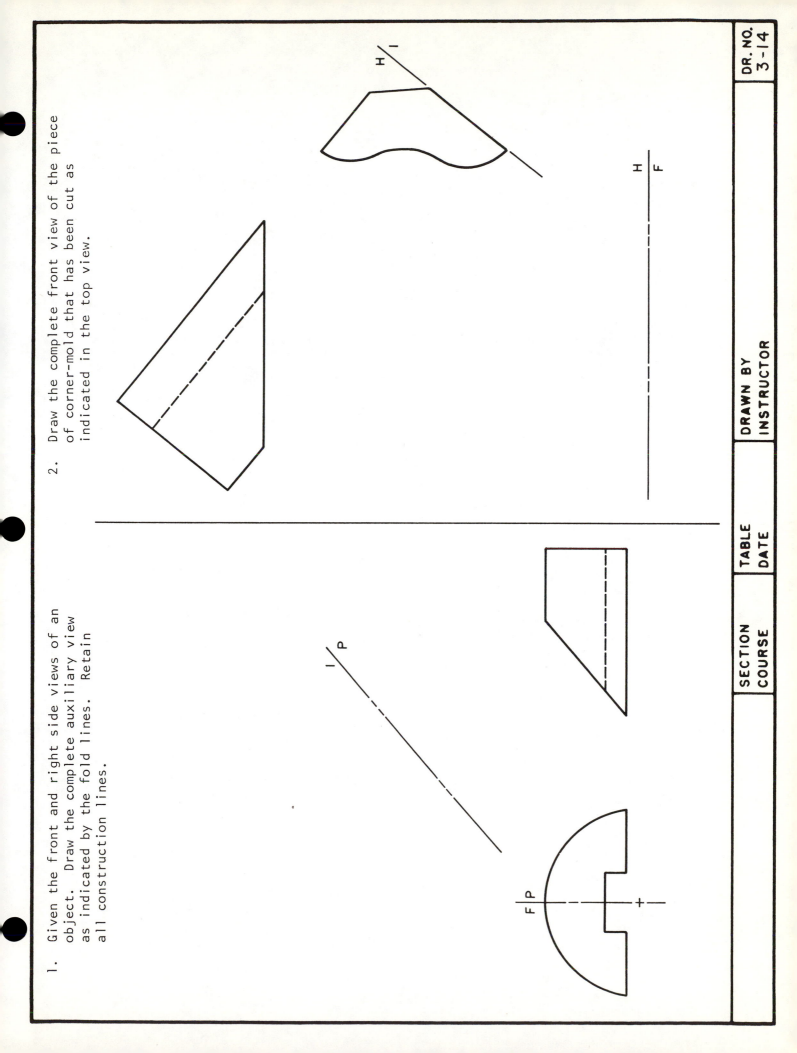

1. Given the front and right side views of an object. Draw the complete auxiliary view as indicated by the fold lines. Retain all construction lines.

2. Draw the complete front view of the piece of corner-mold that has been cut as indicated in the top view.

Given the pictorial view of an object. Draw the front, top and right side views and a complete
auxiliary view to show the true shape of surface A. Scale: Full size.

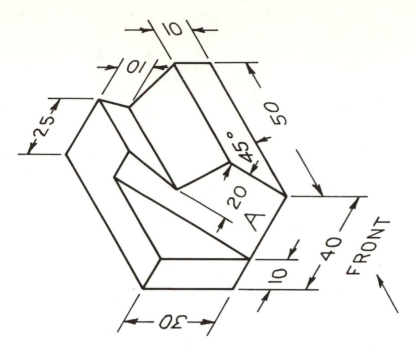

| SECTION | TABLE | DRAWN BY | DR. NO. |
| COURSE | DATE | INSTRUCTOR | 3-15 |

Given the partial front view and the complete right side view of a slider guide. Complete the front view, using the principal of parallelism. Find the true shape of the oblique surface ABC. Draw the complete first auxiliary view and eliminate hidden lines in the second auxiliary view.

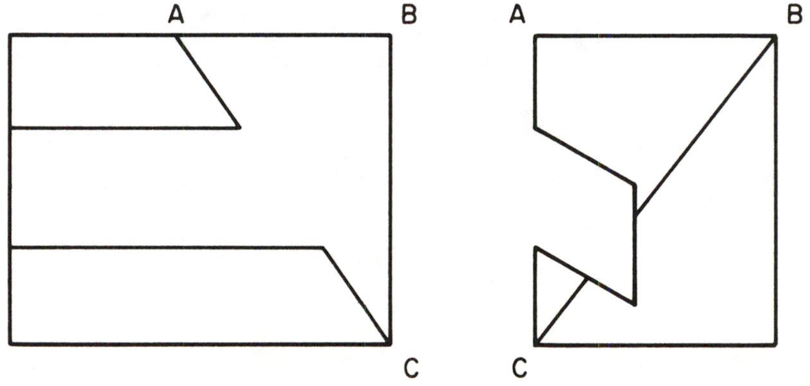

DR. NO.
3-16

SECTION	TABLE	DRAWN BY
COURSE	DATE	INSTRUCTOR

Solve the problems involving parallelism and perpendicularity as indicated. Show necessary construction using only the given views.

1. Construct a plane CDE parallel to line AB.

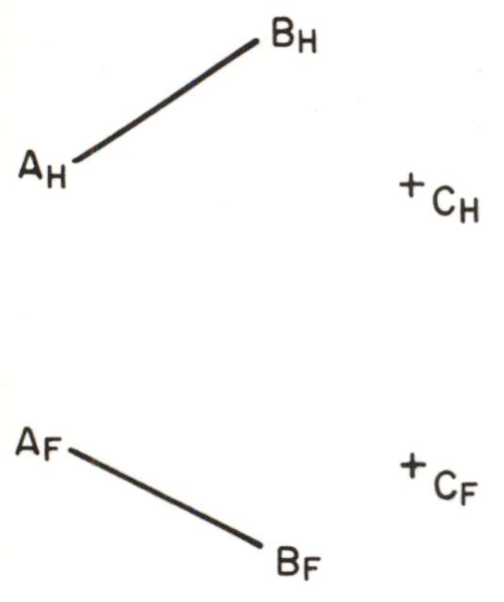

B_H

$+C_H$

A_H

A_F

B_F

$+C_F$

2. Construct a plane DEF parallel to plane ABC.

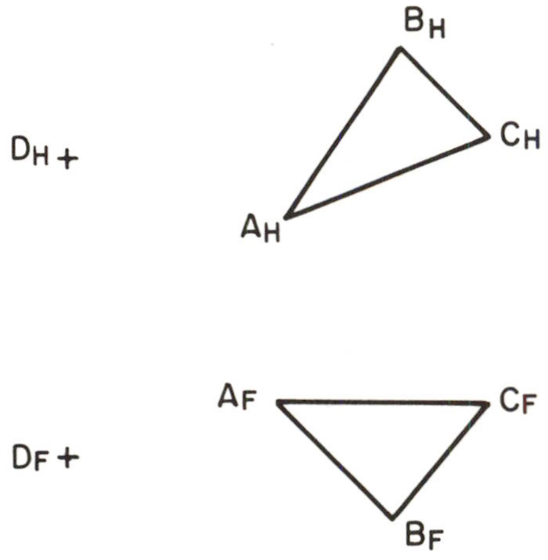

B_H C_H

A_H

$D_H +$

A_F C_F

B_F

$D_F +$

3. Is line AB parallel to plane CDE? Ans. _____

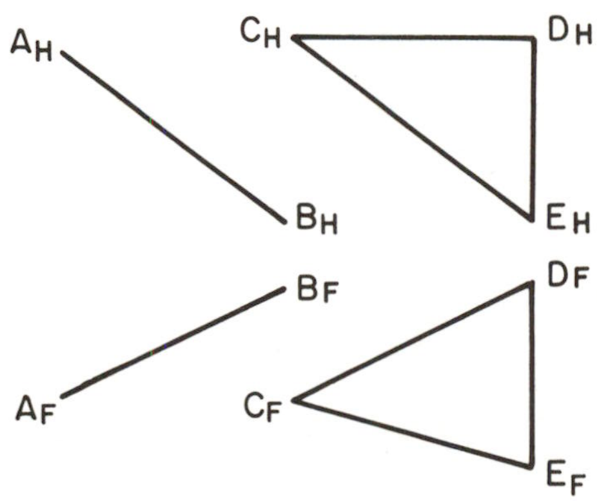

D_H

C_H E_H

B_H

A_H

D_F

E_F

B_F

C_F

A_F

4. Is plane ABC parallel to plane DEF? Ans. _____

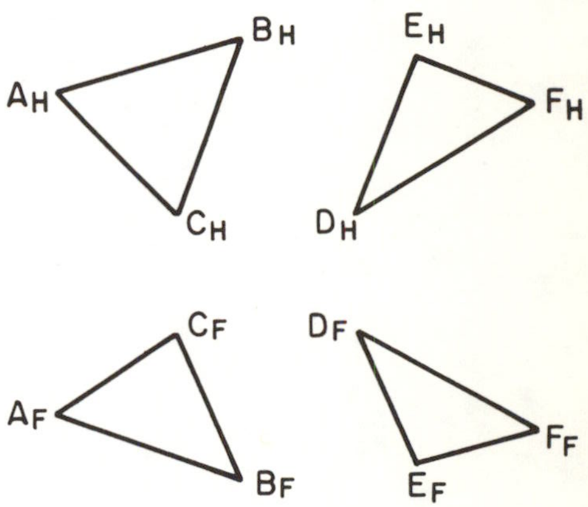

F_H

E_H

D_H

B_H

A_H C_H

F_F

D_F E_F

C_F B_F

A_F

5. Construct a line AB perpendicular to line CD.

D_H

C_H

$+ B_H$

C_F ——— D_F

$+ B_F$

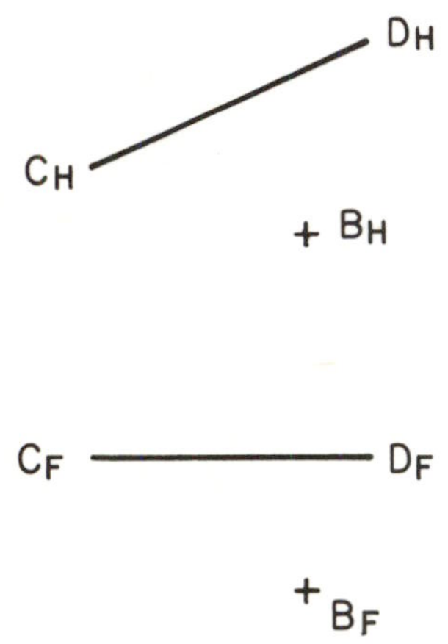

6. Construct a line AB perpendicular to line CD.

$+ B_H$

C_H D_H

C_F

D_F

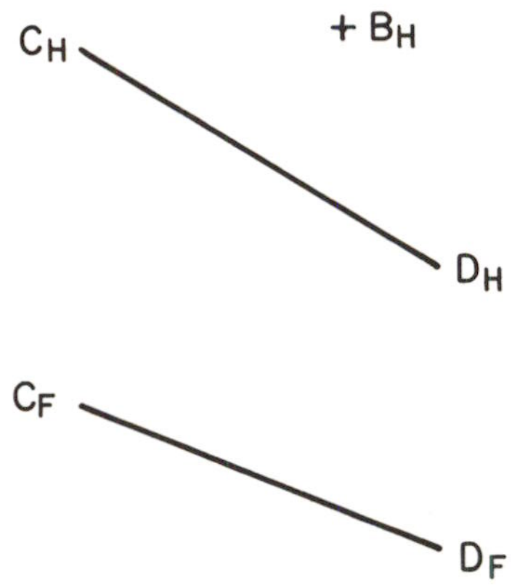

7. Construct a line OP perpendicular to plane AOBC at point O.

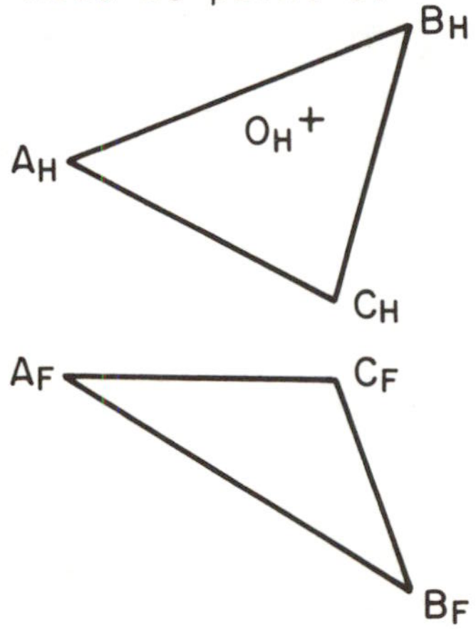

B_H

$O_H +$ C_H

A_H

C_F B_F

A_F

8. Construct a line OP through point O perpendicular to plane ABC.

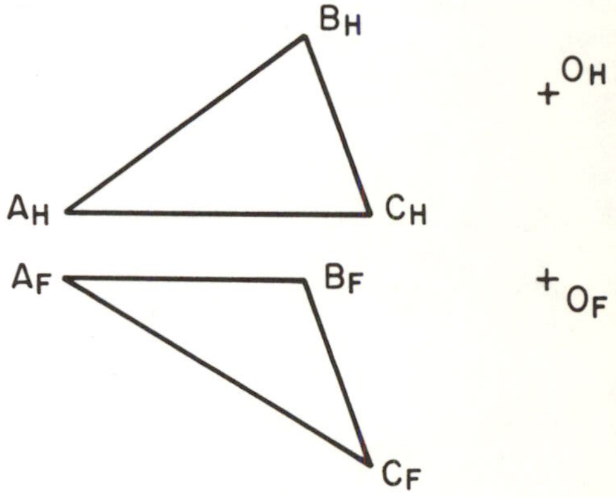

$+ O_H$

$+ O_F$

B_H C_H

A_H

B_F C_F

A_F

SECTION	TABLE	DRAWN BY	DR. NO.
COURSE	DATE	INSTRUCTOR	4-1

1. Draw visible and hidden outlines to complete the given views.

TWO CYLINDERS

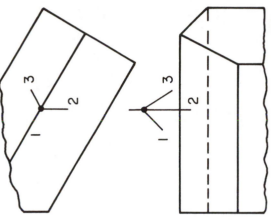

TWO INTERSECTING PLANES

TETRAHEDRON

2. In all views, determine the points at which the guy wires are attached to the roof. Use the auxiliary view method.

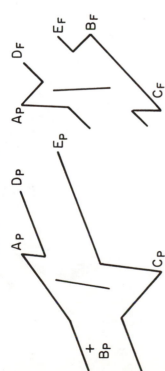

3. Find the piercing points of the control cable with the bulkhead surfaces. Use the cutting plane method and show visibility.

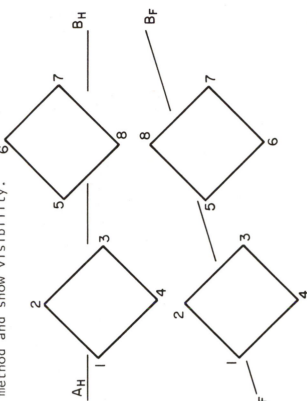

Determine the center of an aperture to be cut in baffle plate ABCD of a test tunnel to permit passage of a particle moving in the direction indicated by the arrow. Use the cutting plane method. Show correct visibility.

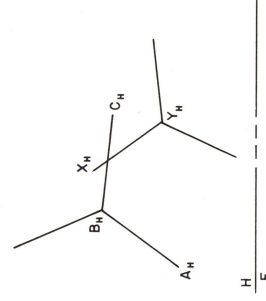

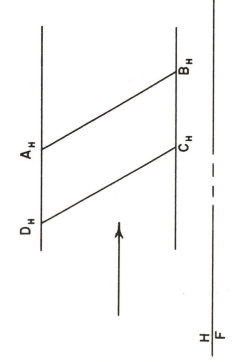

How high above ground level is a straight horizontal cable connecting guys AB, BC, and XY of the radio towers? Draw the horizontal connecting cable in all views. Scale: 1:500

Connecting cable
Ht. above ground = _____

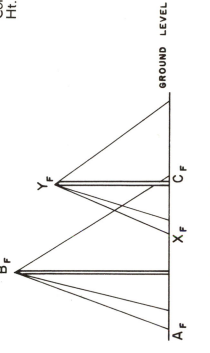

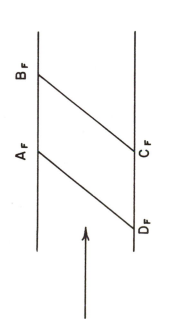

GROUND LEVEL

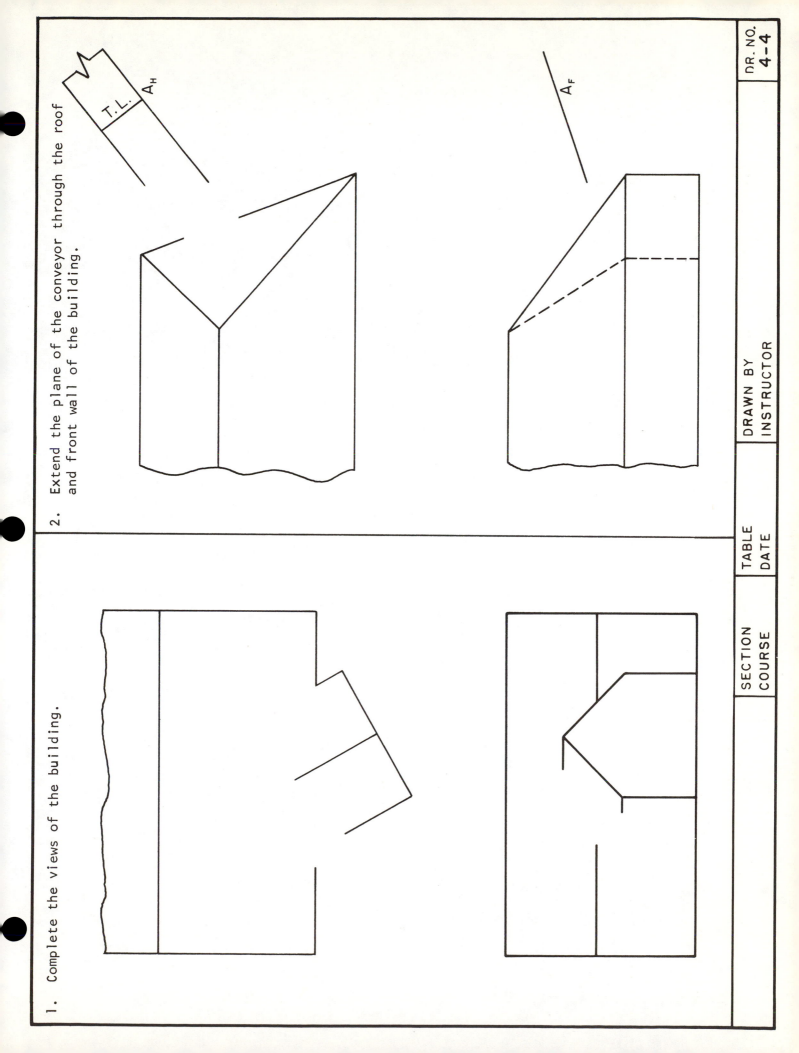

1. Complete the views of the building.

2. Extend the plane of the conveyor through the roof and front wall of the building.

T.L.

A_H

A_F

Solve the following problems by rotation. Show and symbolize all construction used in the problem solutions.

1. Find the TL of the line AB.

ANS
_____

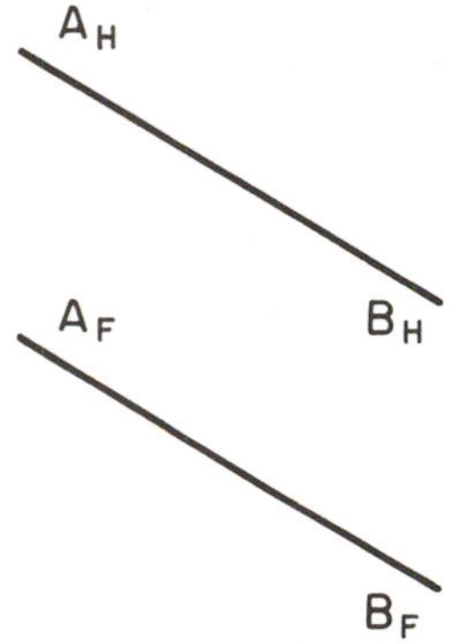

A_H B_H

A_F B_F

2. Find the angle between the line CD and the frontal plane.

ANS
_____

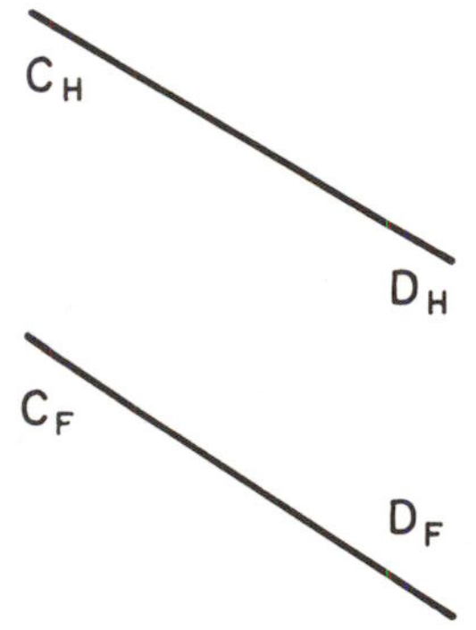

C_H D_H

C_F D_F

3. Find the angle between the line EF and the horizontal plane.

ANS
_____

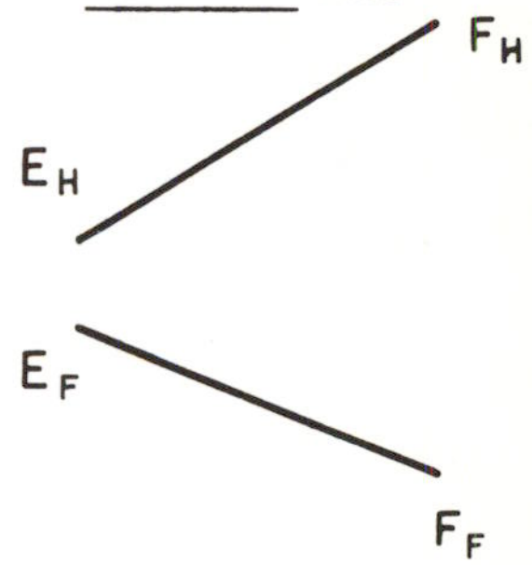

F_H

E_H

E_F F_F

4. Find the angle between the line AB and the profile plane.

ANS
_____

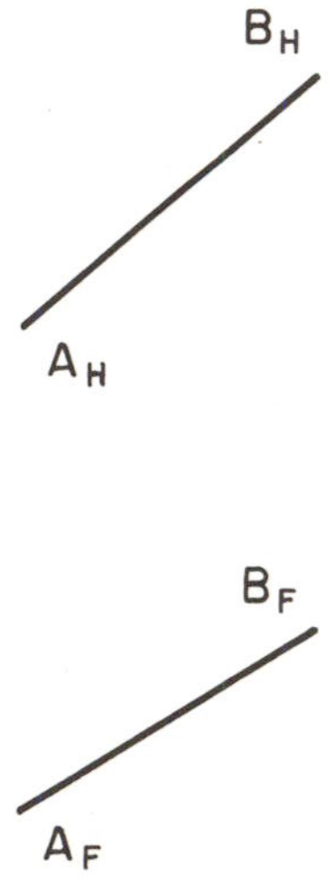

B_H

A_H

B_F

A_F

5. Find the EV and TS of plane DEF.

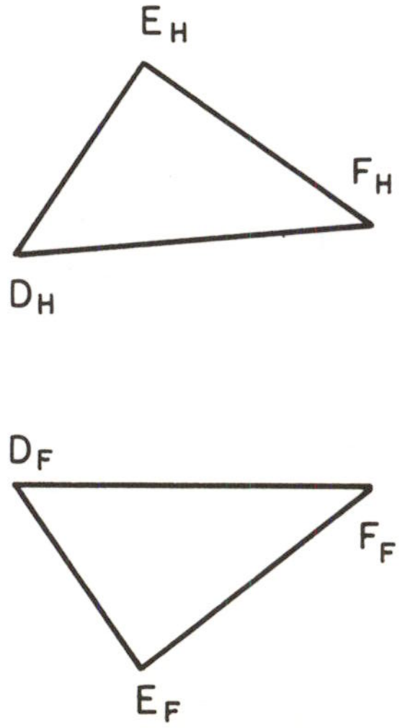

F_H

E_H

D_H

D_F F_F

E_F

6. Rotate the line 3-4 about axis 1-2 until it is parallel to the horizontal plane. Complete all views of the line 3-4.

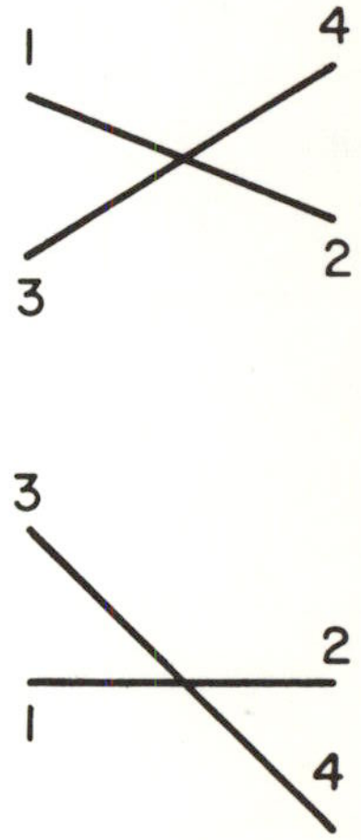

4 2

1 3

2 4

3 1

SECTION   TABLE   DRAWN BY
COURSE   DATE   INSTRUCTOR

DR. NO.
4 - 5

ABCD is the plane of an access door on the inclined side of a hopper. Determine the maximum vertical drop, in centimeters, of corner B as the door is opened about hinge CD. Through what angle has the door been rotated to this position? Scale: 1:20

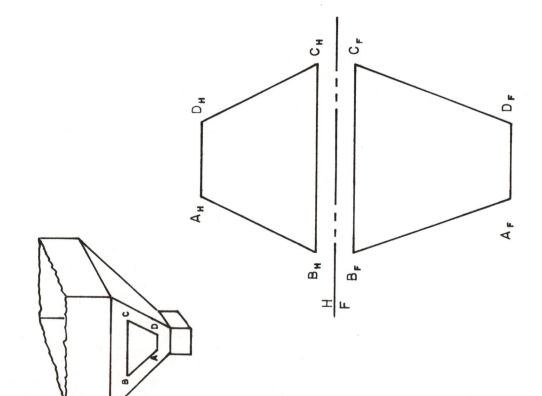

Maximum drop of B = _____ cm

Angle of rotation = _____ °

SECTION	TABLE	DRAWN BY	DR. NO.
COURSE	DATE	INSTRUCTOR	4-6

1. Locate the projections and true lengths of the shortest support members from point A to BC and DE. Label TL lines.

2. Locate the shortest line distance from point C to AB. Solve by constructing a plane through C perpendicular to AB.

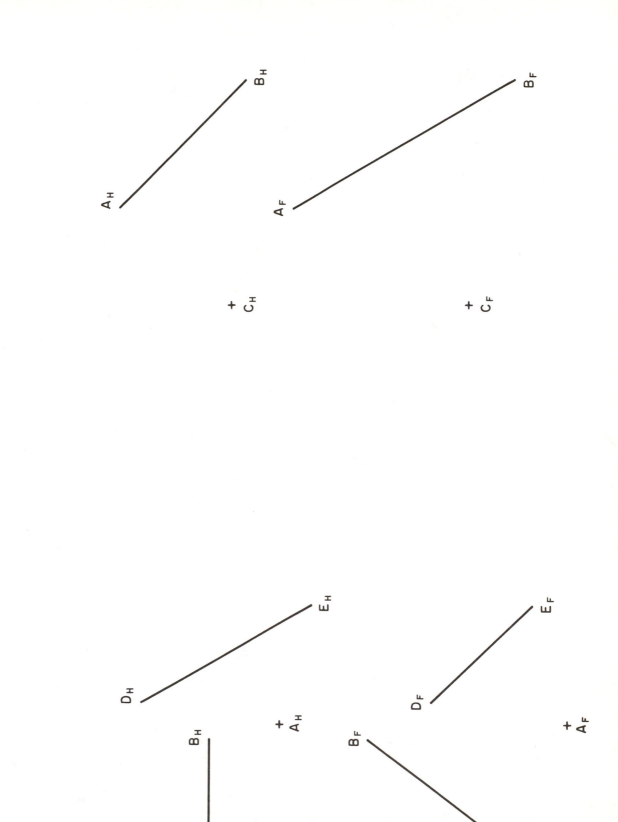

SECTION	TABLE	DRAWN BY
COURSE	DATE	INSTRUCTOR

DR. NO.
4-7

Determine the angles between the struts of the landing gear.

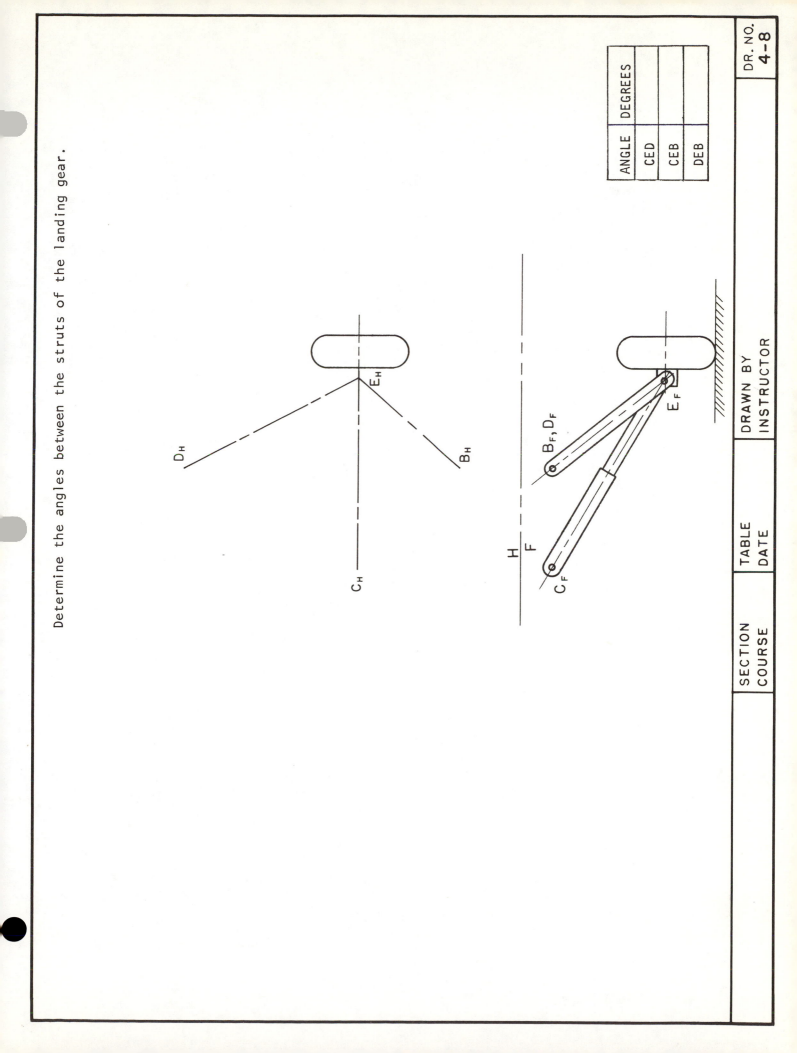

ANGLE	DEGREES
CED	
CEB	
DEB	

| SECTION | TABLE | DRAWN BY | DR. NO. |
| COURSE | DATE | INSTRUCTOR | 4-8 |

Determine the true length, bearing, and grade, slope, or slope angle of line AB; Scale: 1:1. Label TL lines.

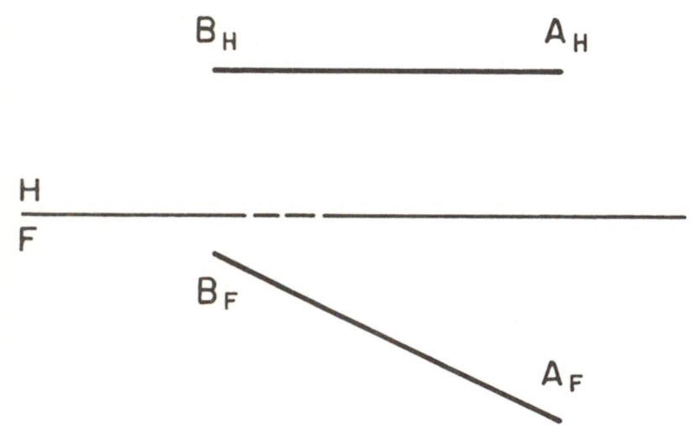

Top-left panel:

A_H — B_H

$\frac{H}{F}$

A_F — B_F

TL _____ Bearing _____ Slope _____

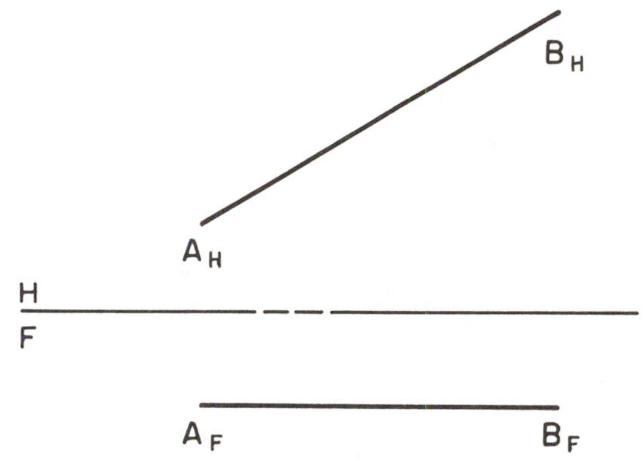

Middle-left panel:

B_H — A_H

$\frac{H}{F}$

B_F — A_F

TL _____ Bearing _____ Slope Angle _____

Bottom-left panel:

A_H — B_H

$\frac{H}{F}$

A_F — B_F

TL _____ Bearing _____ Slope _____

Top-right panel:

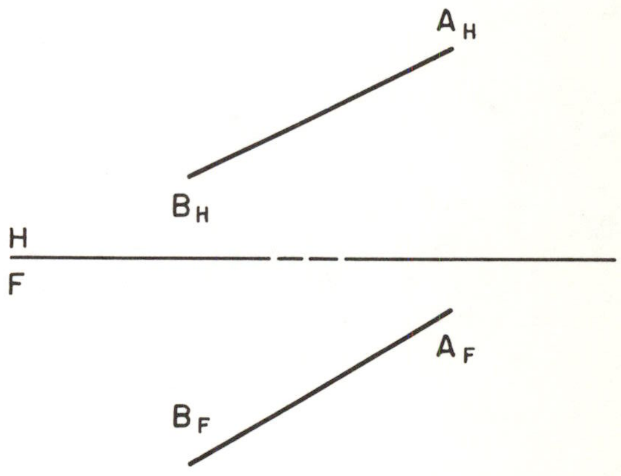

A_H — B_H

$\frac{H}{F}$

A_F — B_F

TL _____ Bearing _____ Grade _____

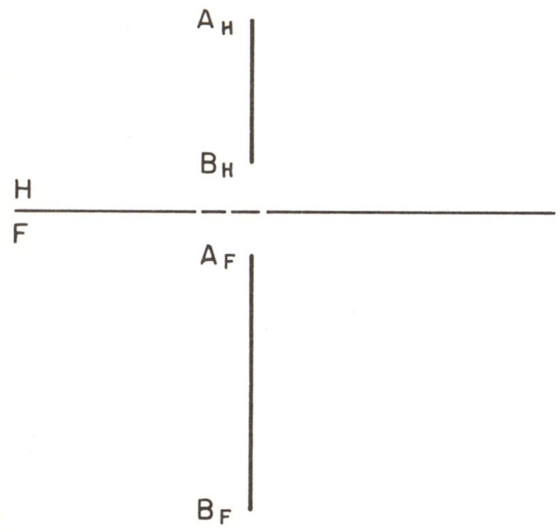

Middle-right panel:

B_P — A_P

$\frac{F}{P}$

B_F — A_F

TL _____ Bearing _____ Grade _____

Bottom-right panel:

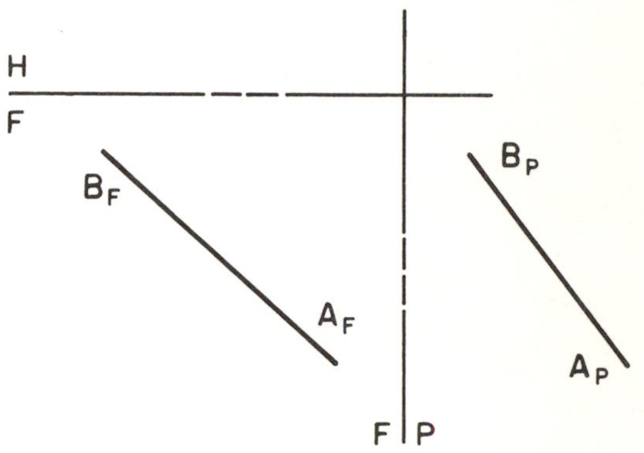

B_P — A_P

$\frac{F}{P}$

A_F — B_F

TL _____ Bearing _____ Slope _____

A_H — B_H

$\frac{H}{F}$

A_F — B_F

TL _____ Bearing _____ Slope Angle _____

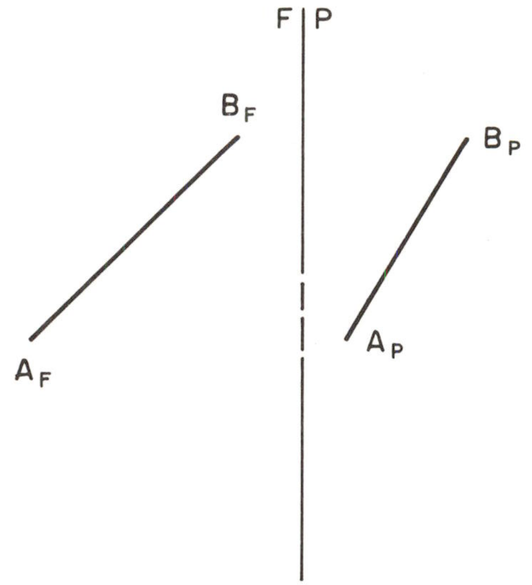

DR. NO. 4-9

DRAWN BY INSTRUCTOR

SECTION _____   TABLE _____
COURSE _____   DATE _____

A particle traveling in the direction shown will strike plane 1 2 3 4,. The angle of reflection of the particle from the plane equals its angle of incidence. Determine the path of the reflected particle and show this path in all views.

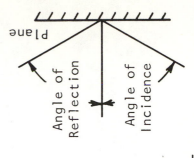

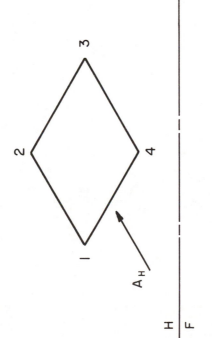

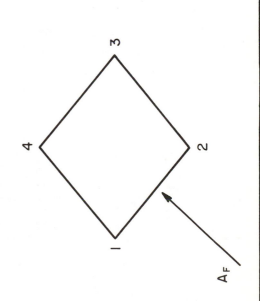

Angle of incidence = _____

| SECTION | TABLE | DRAWN BY | DR. NO. |
| COURSE | DATE | INSTRUCTOR | 4-I0 |

1. Determine the angles that plane DEF
   makes with the two principal planes
   F and P.
   Angle with F _____ .
   Angle with P _____ .

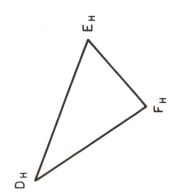

2. Vertical drill holes have located the upper
   face of an ore vein at three points A, B,
   and C. The lower face of the vein was
   located at a point D, 5 m below C. How thick
   is the ore vein? _____ m
   What is the strike and dip of the vein?
   Strike _____ Dip _____
   Scale: 1:400

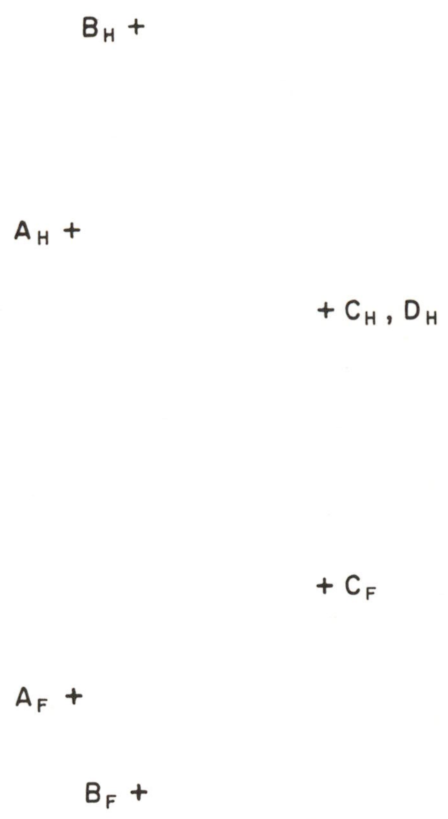

B_H +

A_H +

+ C_H, D_H

+ C_F

A_F +

B_F +

E_H

F_H

D_H

F_F

E_F

D_F

| SECTION | TABLE | DRAWN BY | DR. NO. |
| COURSE | DATE | INSTRUCTOR | 4 – 11 |

1. Determine the angle of bend of the special angle required for the riveted corner of the sheet metal hopper.

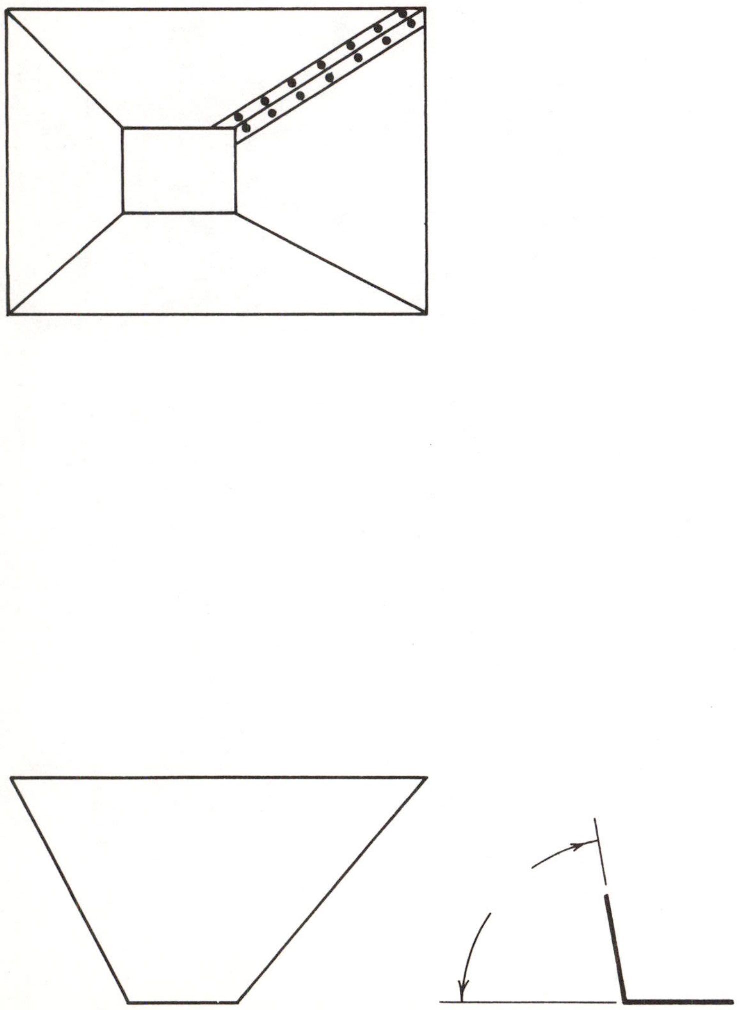

2. Find and mark the angle θ between the faces ABC and ABD of the thread cutting tool.

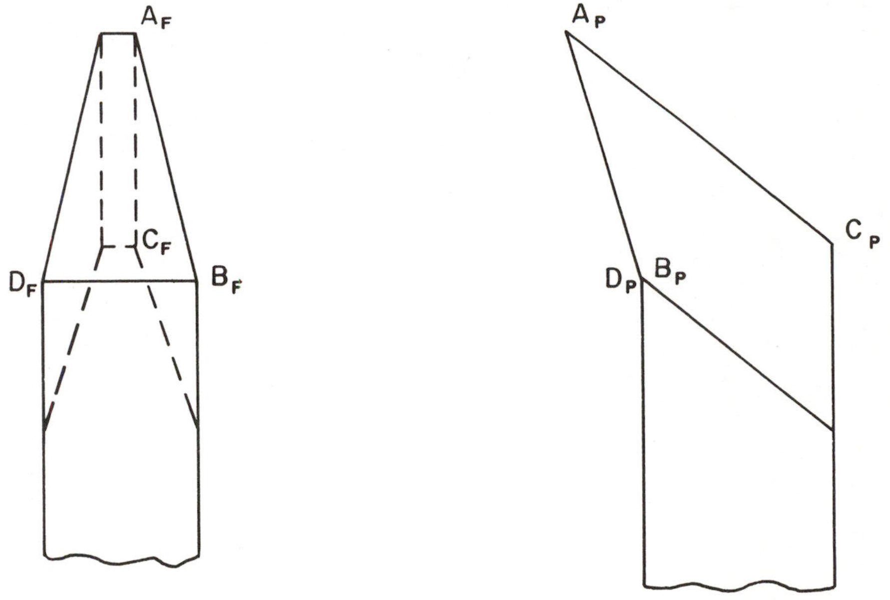

SECTION	TABLE	DRAWN BY	DR. NO.
COURSE	DATE	INSTRUCTOR	4-12

Determine the clearance between the closest guys of the radio
antenna towers. Scale: 1:2000

Show the projection of the clearance distance on all views.

Clearance Distance = _____

H
F

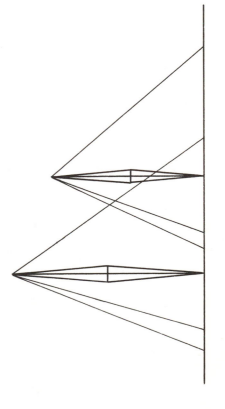

SECTION   TABLE   DRAWN BY
COURSE   DATE   INSTRUCTOR

DR. NO.
4-13

1. The minimum distance allowable between the center lines, AB and CD, of two electric cables is 2m. Determine and dimension the shortest distance between the center lines. If this distance is less than 2m, relocate cable DC by raising point D vertically so as to satisfy the minimum distance requirement.
Scale: 1:100

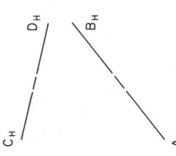

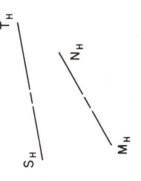

2. MN and ST are the center lines of two pipe lines which are to be connected by an 8m long line AB perpendicular to line MN. Find the top and front views and true length of AB. Does your solution show the true angle between AB and ST?
Scale: 1:400

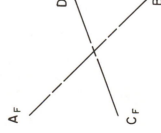

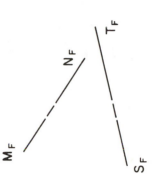

DR. NO.
4-14

DRAWN BY
INSTRUCTOR

SECTION   TABLE
COURSE   DATE

Given the location of the center lines of two structural members. Find the true lengths and projections of the vertical connector and the shortest horizontal connector from AB to CD. Scale: 1:400.

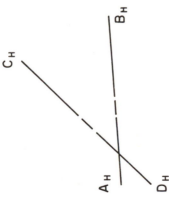

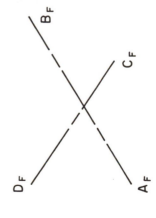

| SECTION | TABLE | DRAWN BY |
| COURSE | DATE | INSTRUCTOR |

DR. NO.
4-15

Given the location of the center lines of two mine shafts. Find the true length of the shortest +35% grade connector from XY to ST. Show this connector in all views. Scale: 1:500.

True length of shortest +35% connector = _____

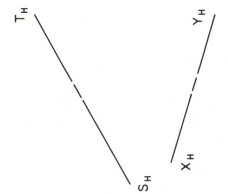

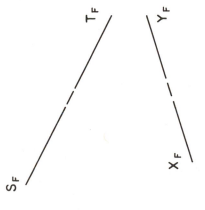

SECTION	TABLE	DRAWN BY
COURSE	DATE	INSTRUCTOR

DR. NO.
4-16

Given top, auxiliary and partial front views of the rectangular prism, EFGH, intersecting the triangular prism, ABC.

Complete the front view showing the line of intersection of the two prisms. SHOW ALL CONSTRUCTION. LABEL ALL POINTS ON THE INTERSECTION.

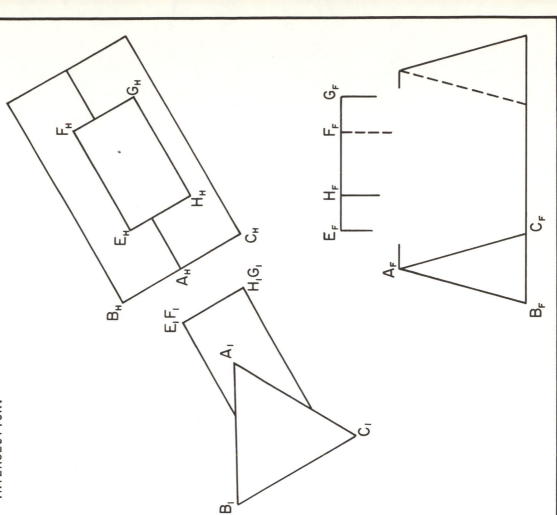

Given the top, left and partial front view of the triangular prism, ABC, intersecting the square prism DEFG.

Complete the front view showing the line of intersection of the two prisms. SHOW ALL CONSTRUCTION. LABEL ALL POINTS ON THE INTERSECTION.

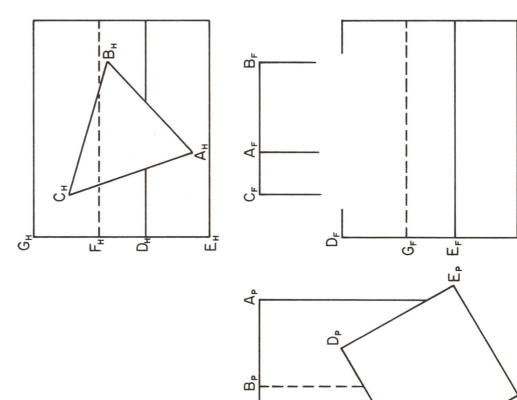

SECTION	TABLE	DRAWN BY	DR. NO.
COURSE	DATE	INSTRUCTOR	5-1

Complete the views of the two intersecting cylindrical surfaces. Identify the cutting planes.

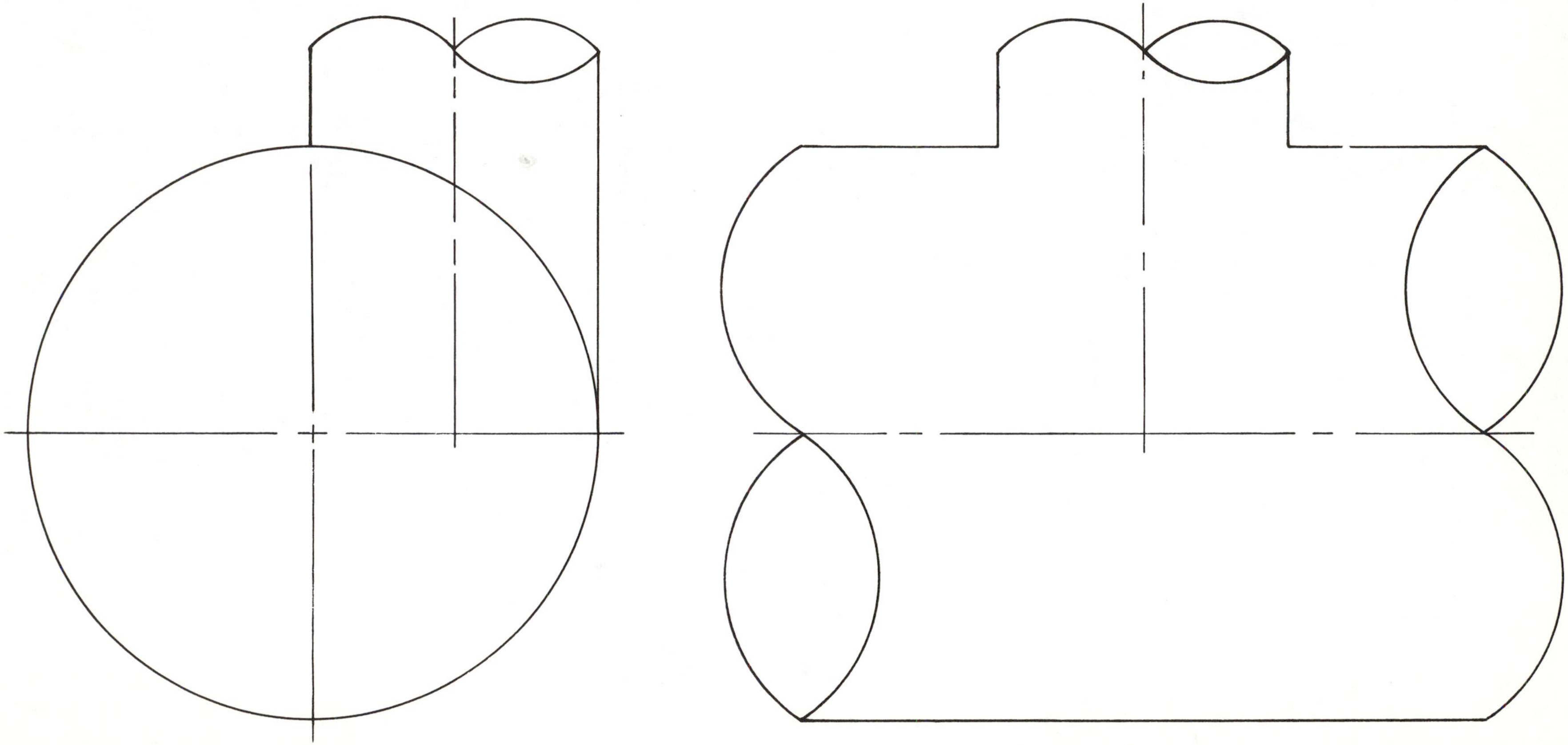

SECTION   TABLE   DRAWN BY
COURSE   DATE   INSTRUCTOR

DR. NO.
5-2

Complete the views of the cylindrical tank and the intersecting cylindrical pipe. Identify the cutting planes.

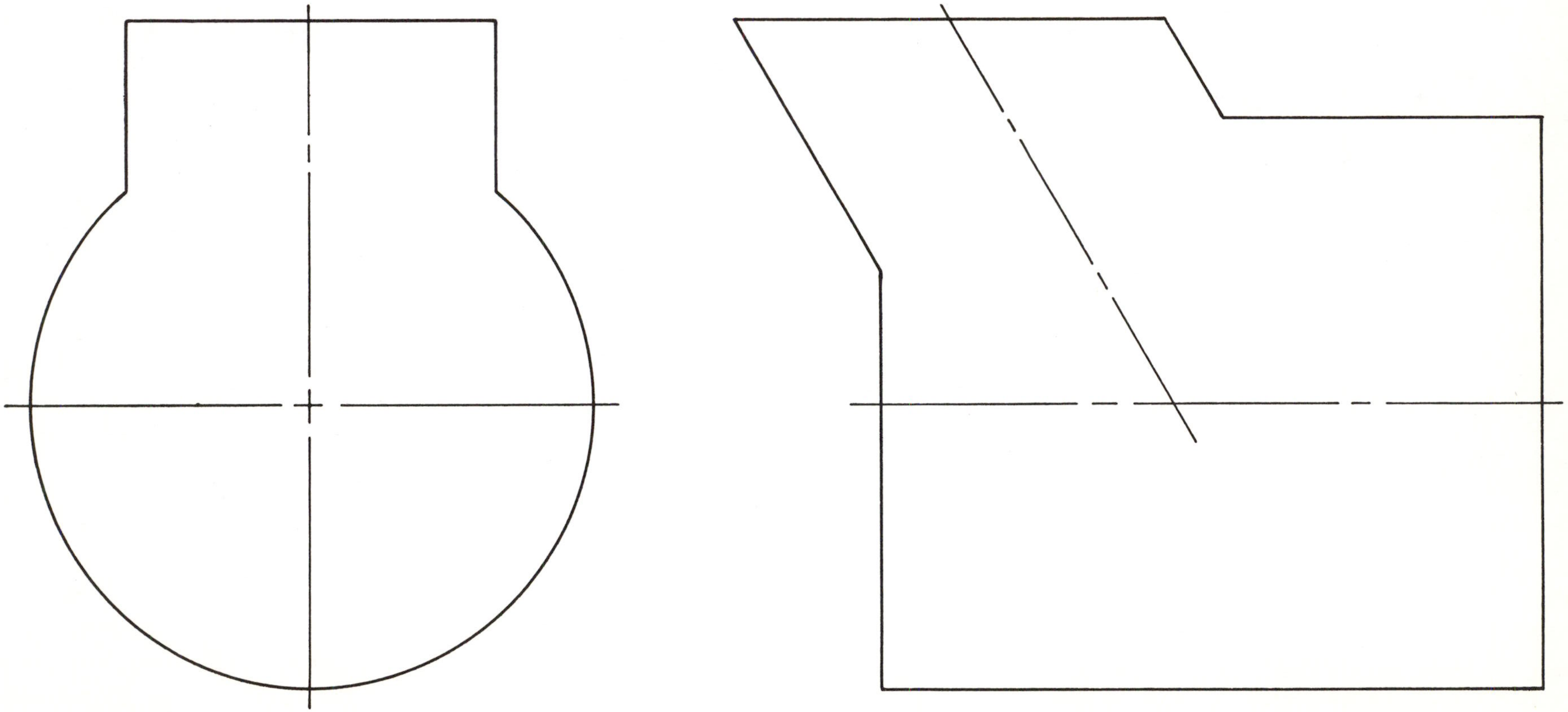

DR. NO.
5-3

SECTION
COURSE

TABLE
DATE

DRAWN BY
INSTRUCTOR

Determine the intersections of the cylindrical supply duct and the square discharge outlet with the right circular conical hopper.

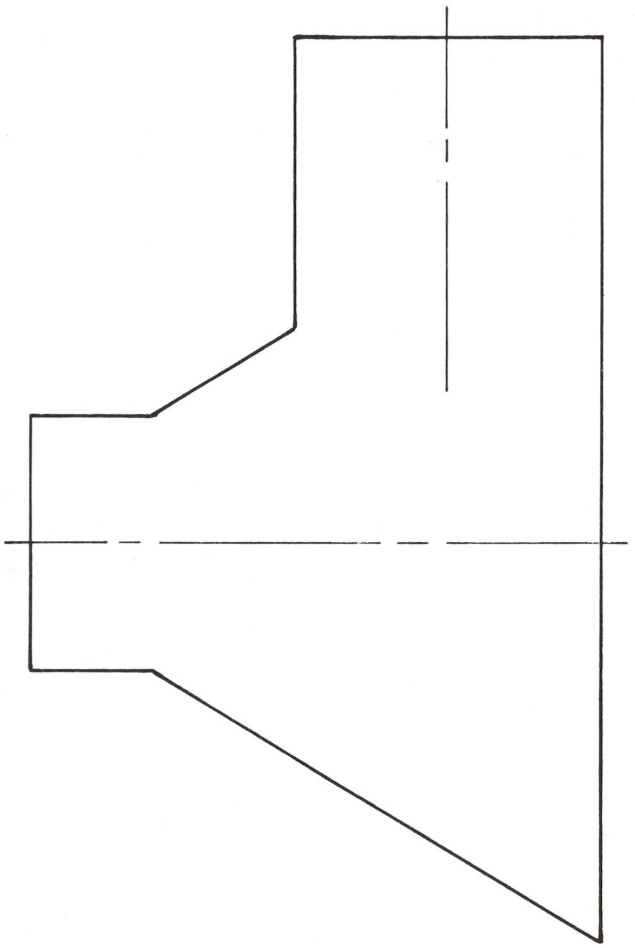

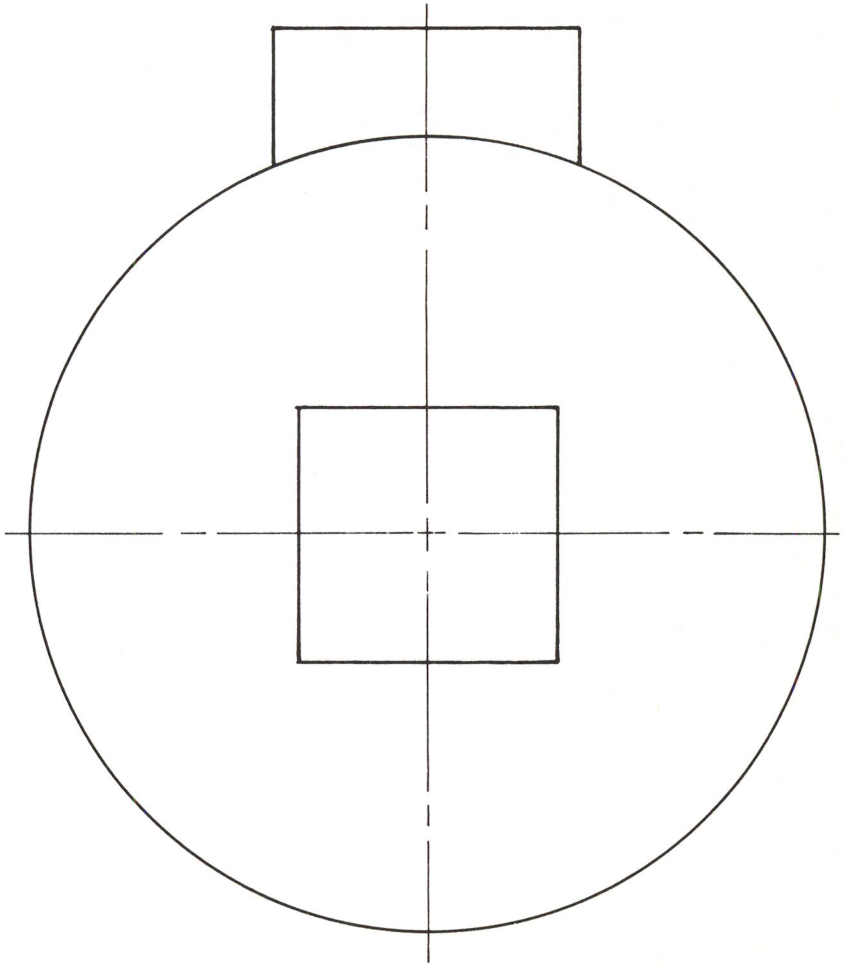

TOP

FRONT

SECTION
COURSE

TABLE
DATE

DRAWN BY
INSTRUCTOR

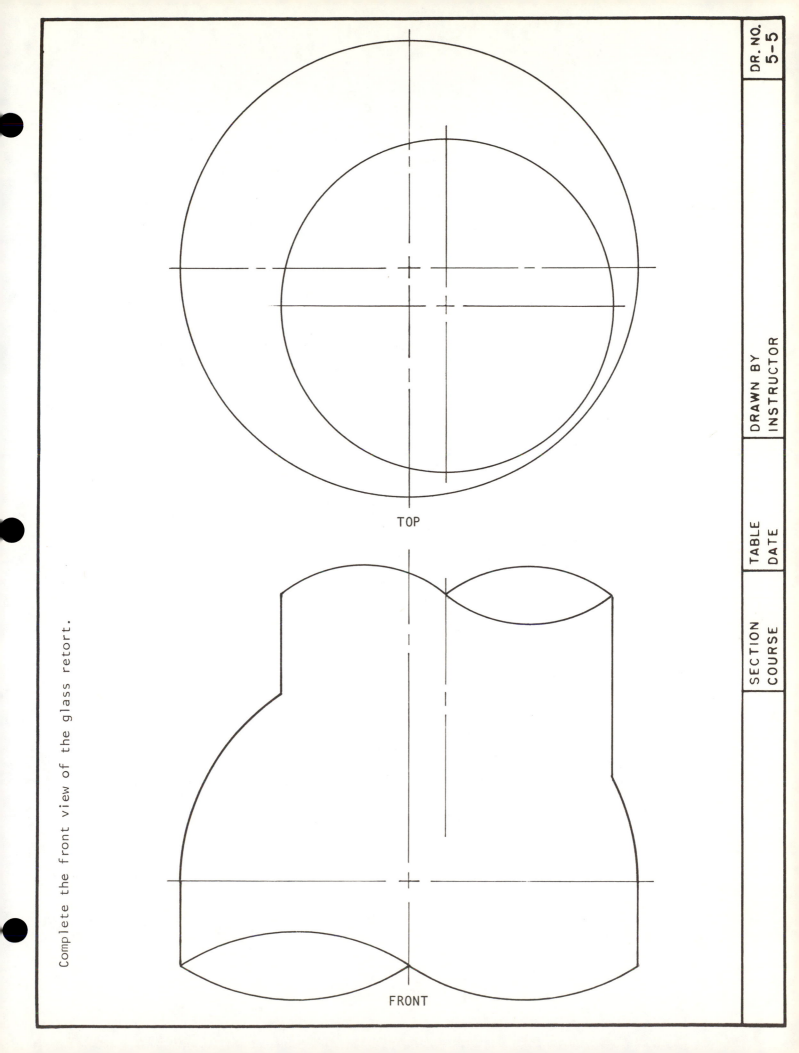

Complete the front view of the glass retort.

TOP

FRONT

DRAWN BY
INSTRUCTOR

TABLE
DATE

SECTION
COURSE

Determine the top view and the true shape of the intersection of the right circular cone and cutting plane A-A.
What is the name of this curve of intersection? _____

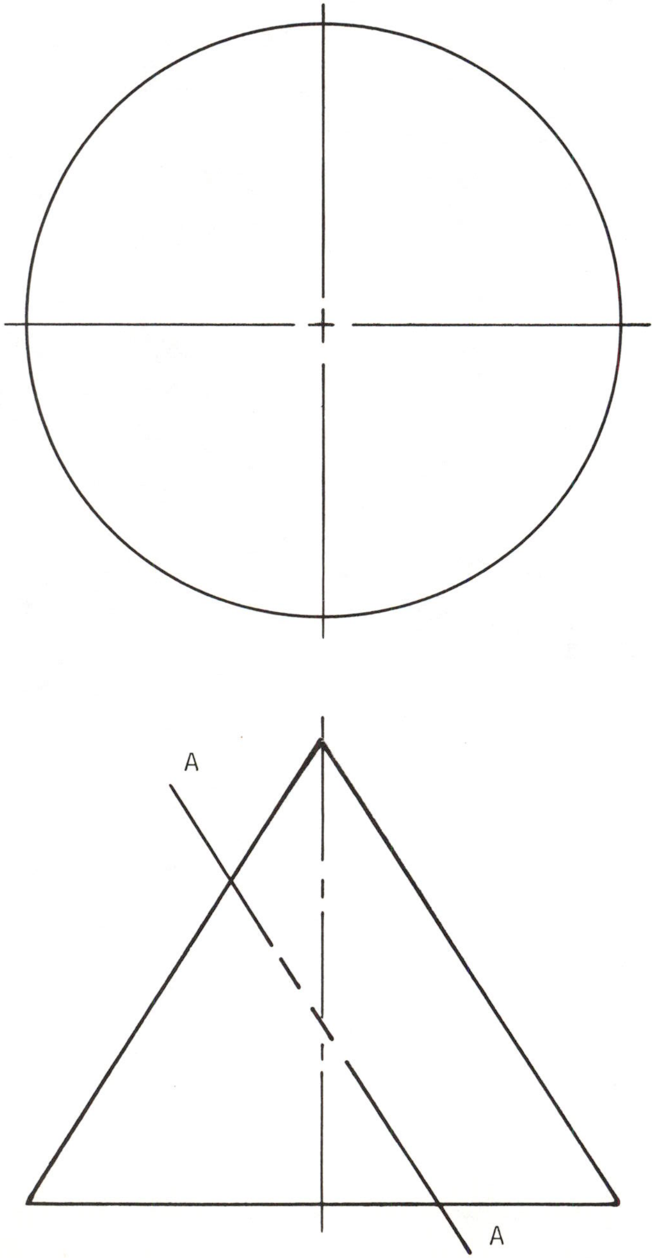

DR. NO.
5-6

DRAWN BY
INSTRUCTOR

SECTION   TABLE
COURSE   DATE

Develop the vertical surfaces of the cylindrical ensilage tank and the square ventilator filling hopper. Show the inside surface up and make the seams as short as possible.

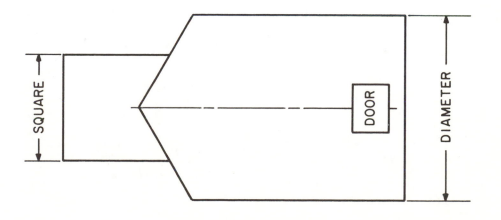

SQUARE

DOOR

DIAMETER

| SECTION | TABLE | DRAWN BY | DR. NO. |
| COURSE | DATE | INSTRUCTOR | 5-7 |

Complete the front view and develop a symmetrical half
of the right circular cone transition piece.

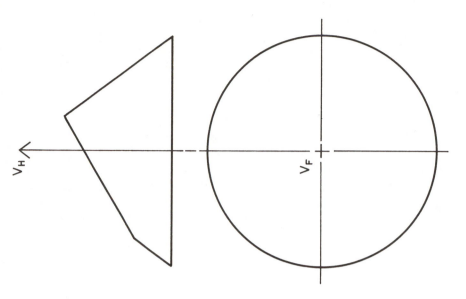

Develop a symmetrical half of the pyramidal frustum
grain bin. Start the development with the seam line
AB and show the inside up.

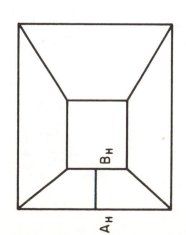

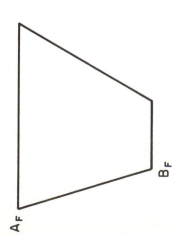

SECTION	TABLE	DRAWN BY	DR. NO.
COURSE	DATE	INSTRUCTOR	5-8

Develop a symmetrical half of the conical transition connection. Show inside up. Begin with element AB.

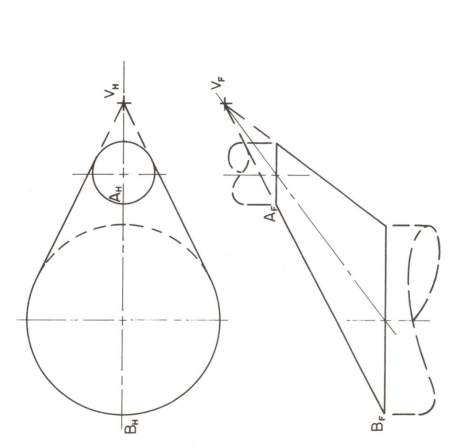

DR. NO.		
5-9		
SECTION	TABLE	DRAWN BY
COURSE	DATE	INSTRUCTOR

Find the true right section and draw the development of the galvanized sheet metal ventilation duct represented by the line drawing below. Determine the true shape of the floor cut out. Start development at the seam AB.

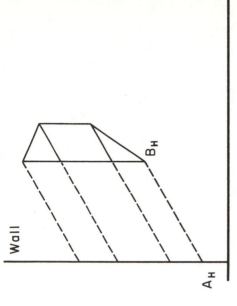

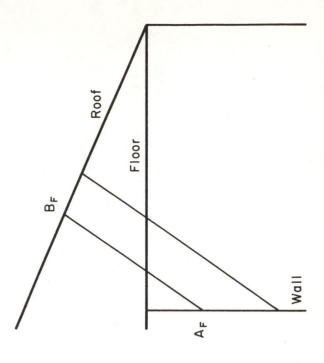

SECTION	TABLE	DRAWN BY	DR. NO.
COURSE	DATE	INSTRUCTOR	5-10

Draw and develop a sheet metal transition piece connecting the two horizontal rectangular ducts. Use triangulation and show inside up.

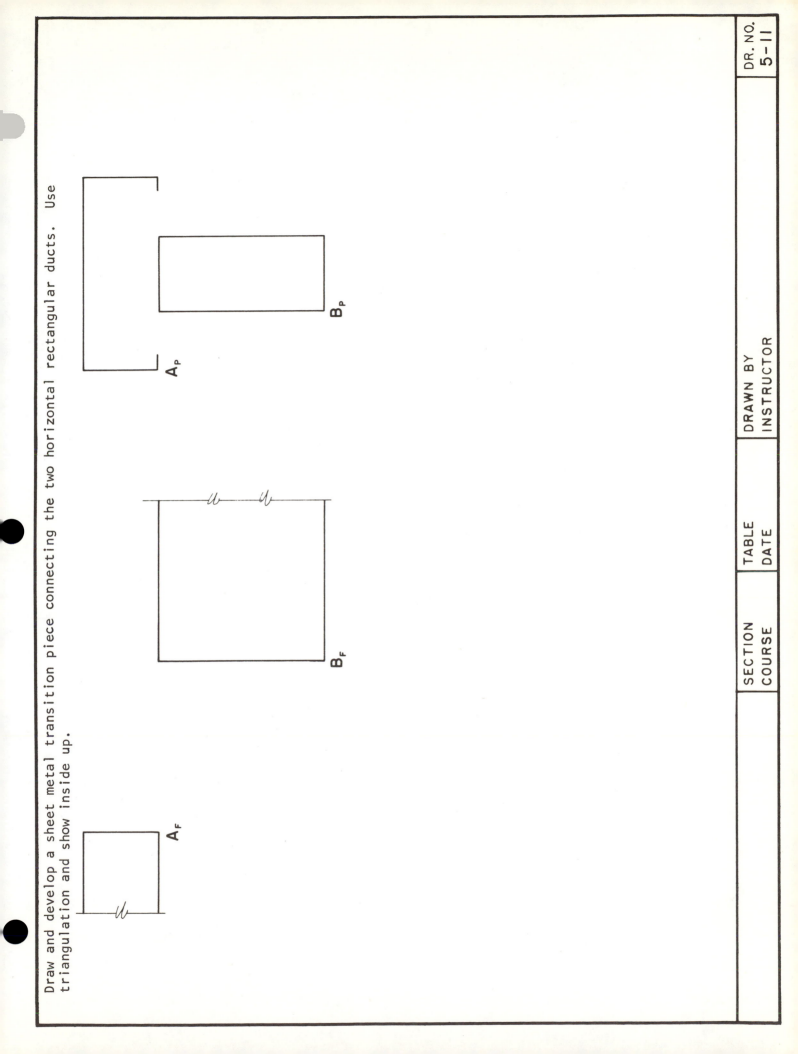

A_F

B_F

A_P

B_P

SECTION
COURSE

TABLE
DATE

DRAWN BY
INSTRUCTOR

DR. NO.
5-11

Draw the indicated section views.

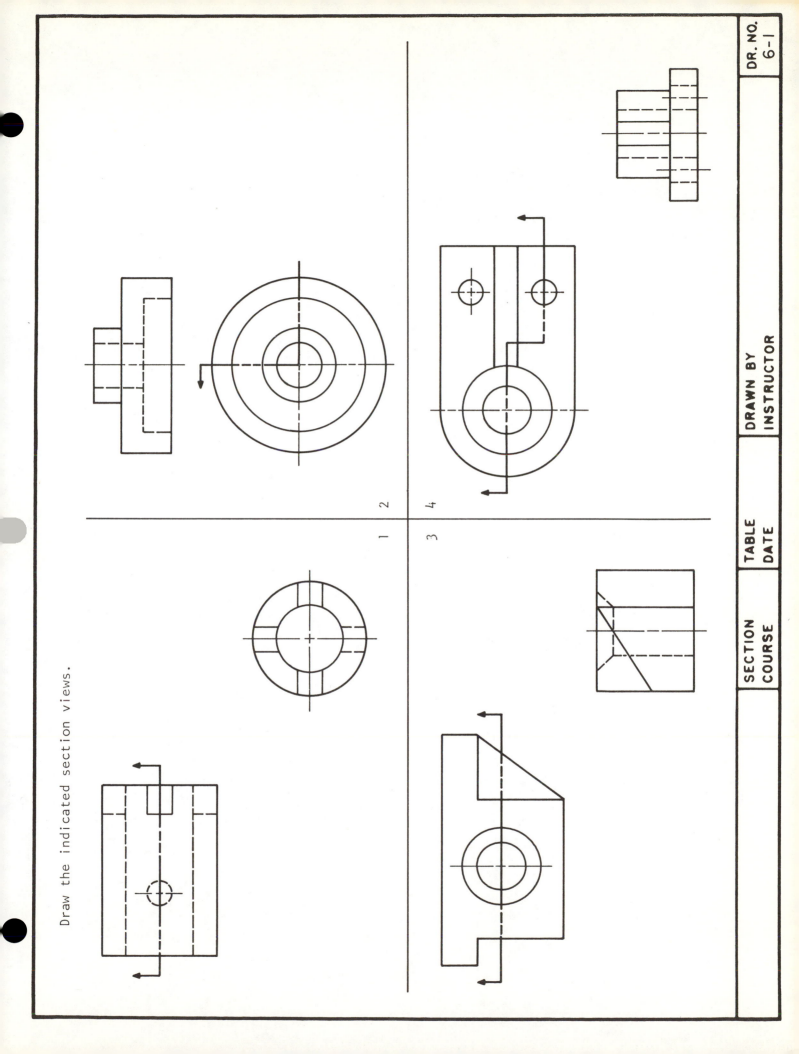

1   2

3   4

SECTION   TABLE   DRAWN BY
COURSE   DATE   INSTRUCTOR

DR. NO.
6-1

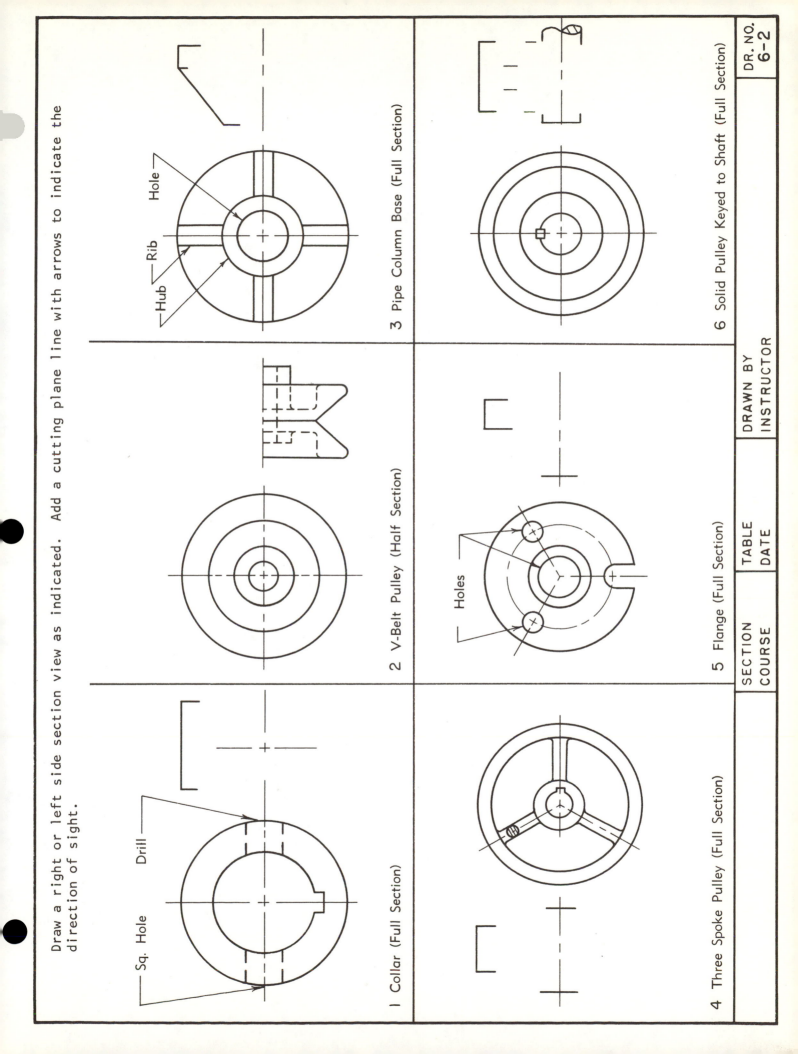

Draw a right or left side section view as indicated. Add a cutting plane line with arrows to indicate the direction of sight.

1 Collar (Full Section)

Sq. Hole

Drill

2 V-Belt Pulley (Half Section)

3 Pipe Column Base (Full Section)

Hole

Rib

Hub

4 Three Spoke Pulley (Full Section)

5 Flange (Full Section)

Holes

6 Solid Pulley Keyed to Shaft (Full Section)

| SECTION | TABLE | DRAWN BY | DR. NO. |
| COURSE | DATE | INSTRUCTOR | 6-2 |

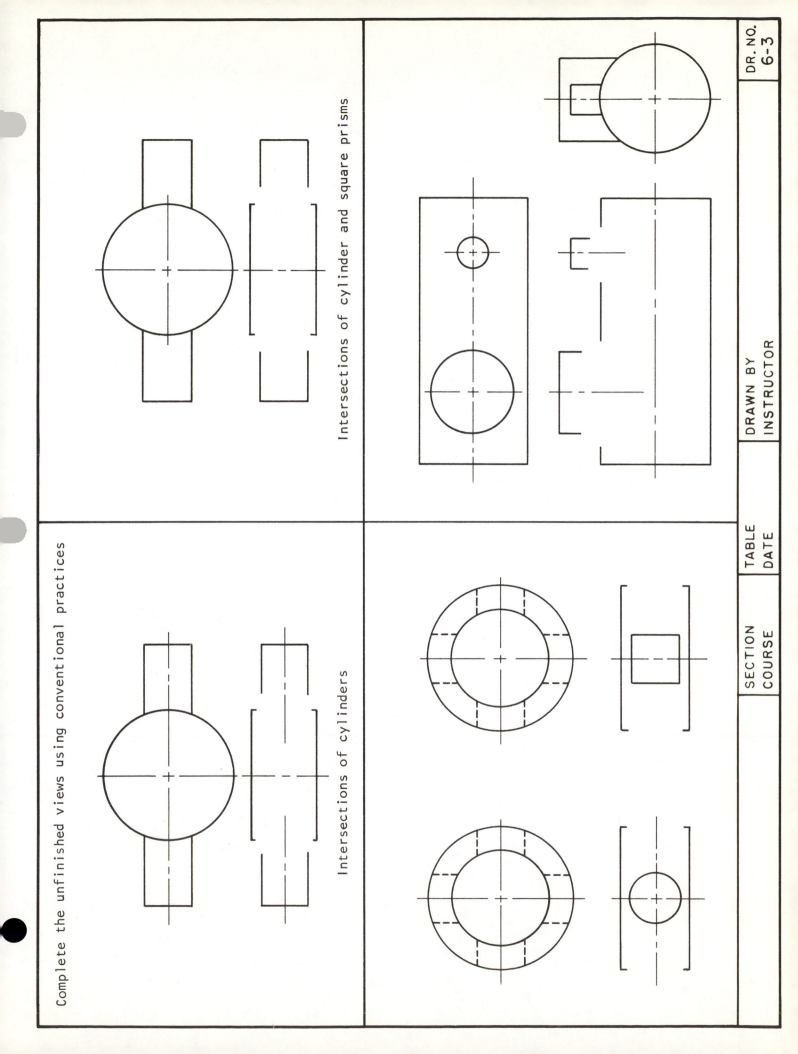

Complete the unfinished views using conventional practices

Intersections of cylinder and square prisms

Intersections of cylinders

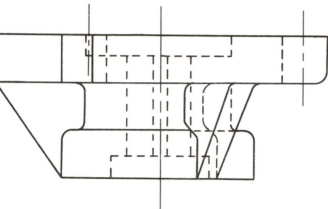

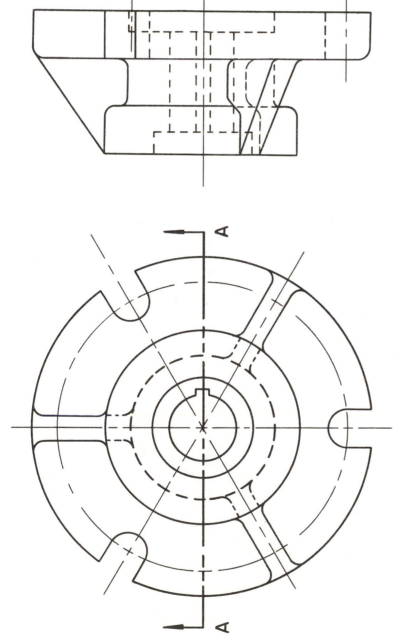

Given the horizontal and true right profile views of a Flanged Pulley. Draw the front view in full section as indicated by the cutting plane line A-A..

| SECTION | TABLE | DRAWN BY | DR. NO. |
| COURSE | DATE | INSTRUCTOR | 6-4 |

Given two views of a machine part. Draw the section view A-A.

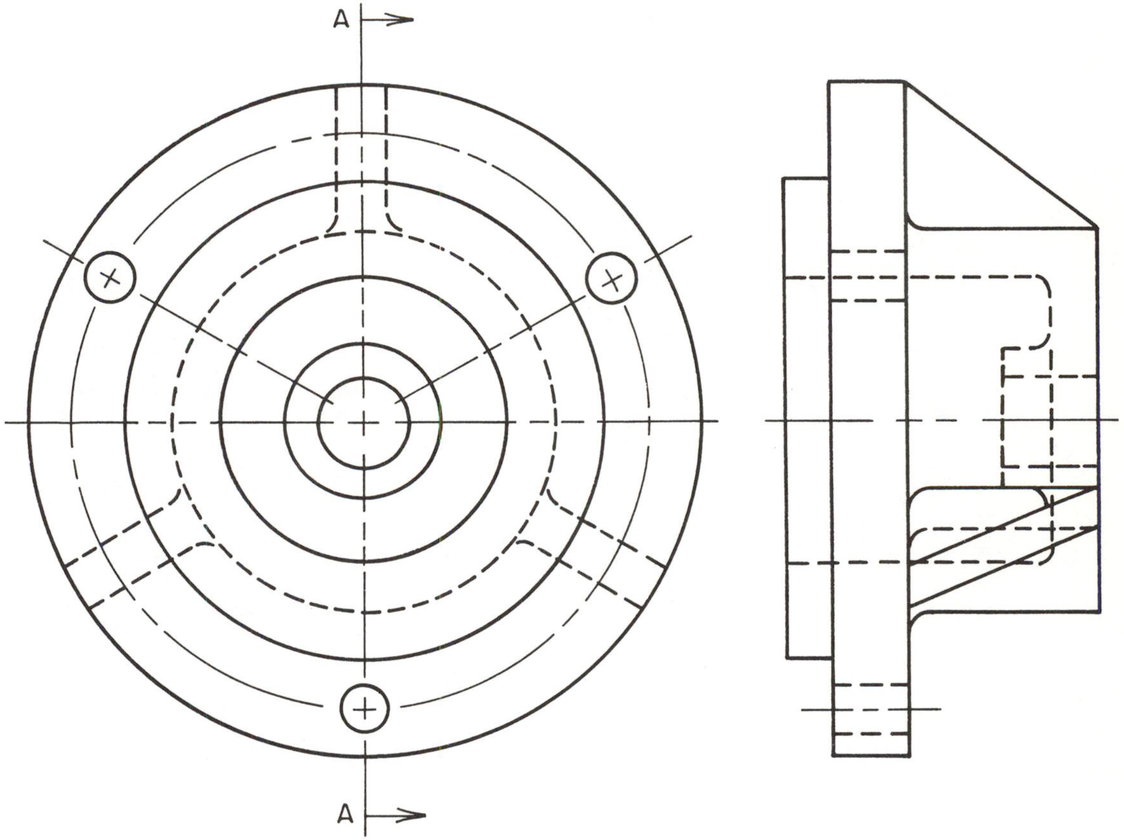

DR. NO.
6-5

DRAWN BY
INSTRUCTOR

TABLE
DATE

SECTION
COURSE

A

A

Add extension lines, dimension lines, leaders, arrow heads and notes to the given views. Use letters S and L to identify size and location dimensions. Do not scale for numerical values.

1. Rectangular and Triangular Prisms

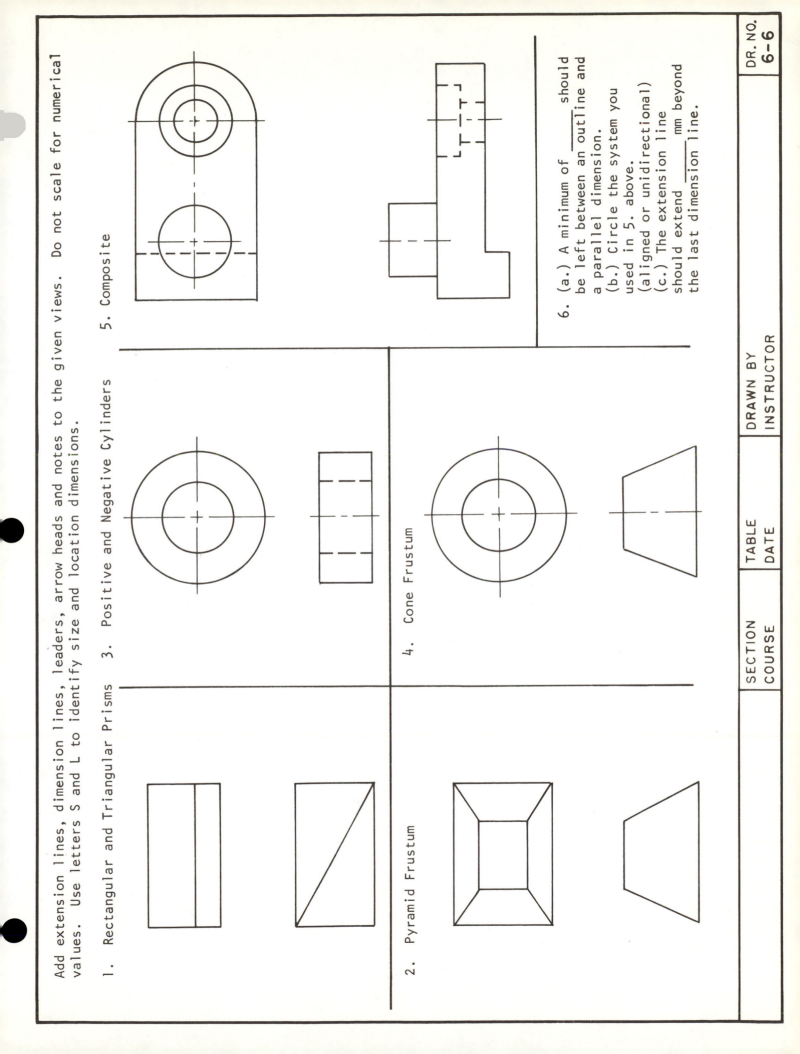

2. Pyramid Frustum

3. Positive and Negative Cylinders

4. Cone Frustum

5. Composite

6. (a.) A minimum of ____ should be left between an outline and a parallel dimension.
(b.) Circle the system you used in 5. above.
(aligned or unidirectional)
(c.) The extension line should extend ____ mm beyond the last dimension line.

| SECTION | TABLE | DRAWN BY |
| COURSE | DATE | INSTRUCTOR |

DR. NO.
6-6

Add extension lines, dimension lines, leaders, arrowheads and notes to the given views. Use the letters S and L to indicate Size and Location dimensions.

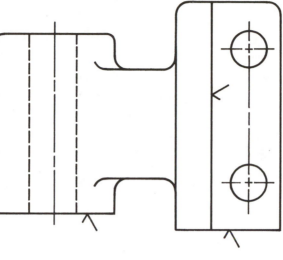

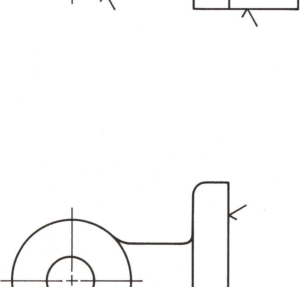

| SECTION | TABLE | DRAWN BY | DR. NO. |
| COURSE | DATE | INSTRUCTOR | 6-7 |

Add extension lines, dimension lines, leaders, arrowheads and notes to the given views. Use the letters S and L to indicate Size and Location dimensions.

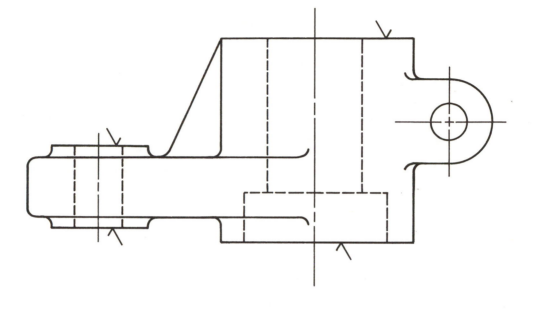

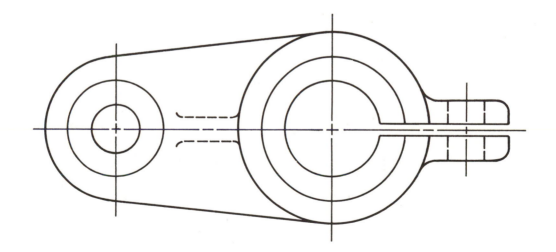

| SECTION | TABLE | DRAWN BY | DR. NO. |
| COURSE | DATE | INSTRUCTOR | 6-8 |

Dimension the drawing of the Shaft Support. Add finish marks and a sub-title or legend for a complete detail working drawing. Scale: Half size

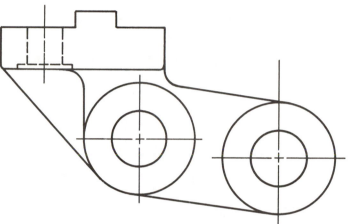

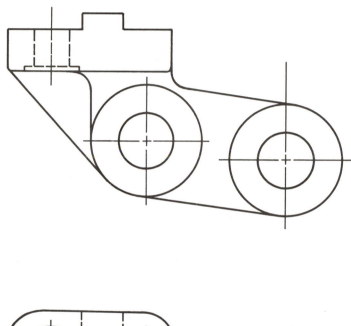

SECTION	TABLE	DRAWN BY
COURSE	DATE	INSTRUCTOR

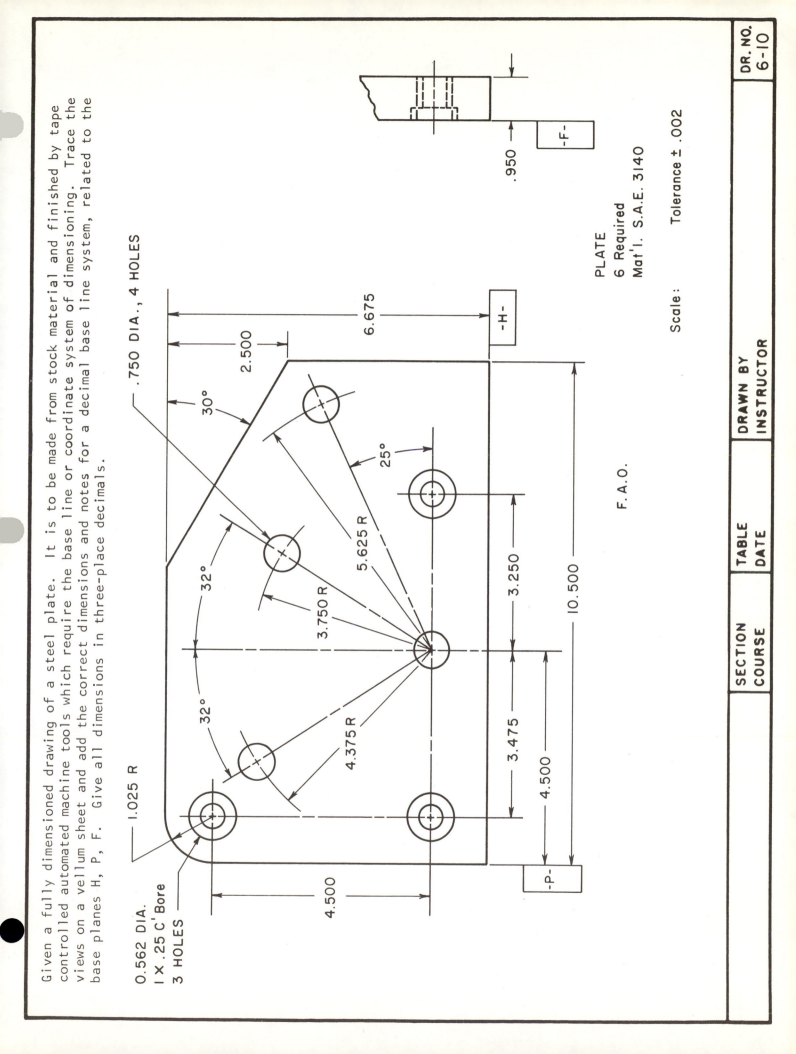

Given a fully dimensioned drawing of a steel plate. It is to be made from stock material and finished by tape controlled automated machine tools which require the base line or coordinate system of dimensioning. Trace the views on a vellum sheet and add the correct dimensions and notes for a decimal base line system, related to the base planes H, P, F. Give all dimensions in three-place decimals.

0.562 DIA.
1 X .25 C' Bore
3 HOLES

1.025 R

.750 DIA., 4 HOLES

2.500

6.675

30°

25°

32°

32°

5.625 R

3.750 R

4.375 R

3.250

10.500

3.475

4.500

4.500

.950

-F-

-H-

-P-

F.A.O.

PLATE
6 Required
Mat'l. S.A.E. 3140

Scale: Tolerance ± .002

| SECTION | TABLE | DRAWN BY | DR. NO. |
| COURSE | DATE | INSTRUCTOR | 6-10 |

Enter one of the following words in each blank space below: "Maximum," "minimum," "allowance," "hole," "shaft."

Hole Diameter — — Tolerance = [Hole Diameter]

[Shaft Diameter] = Tolerance — — Shaft Diameter

Clearance

[Hole Diameter | Basic Size] Basic —— System

[Shaft Diameter | Basic Size] Basic —— System

BASIC PRODUCTION DIMENSIONING SYSTEMS

TOLERANCE is defined as _____

ALLOWANCE is defined as _____

From the given dimensions on each drawing, fill in the blanks below.

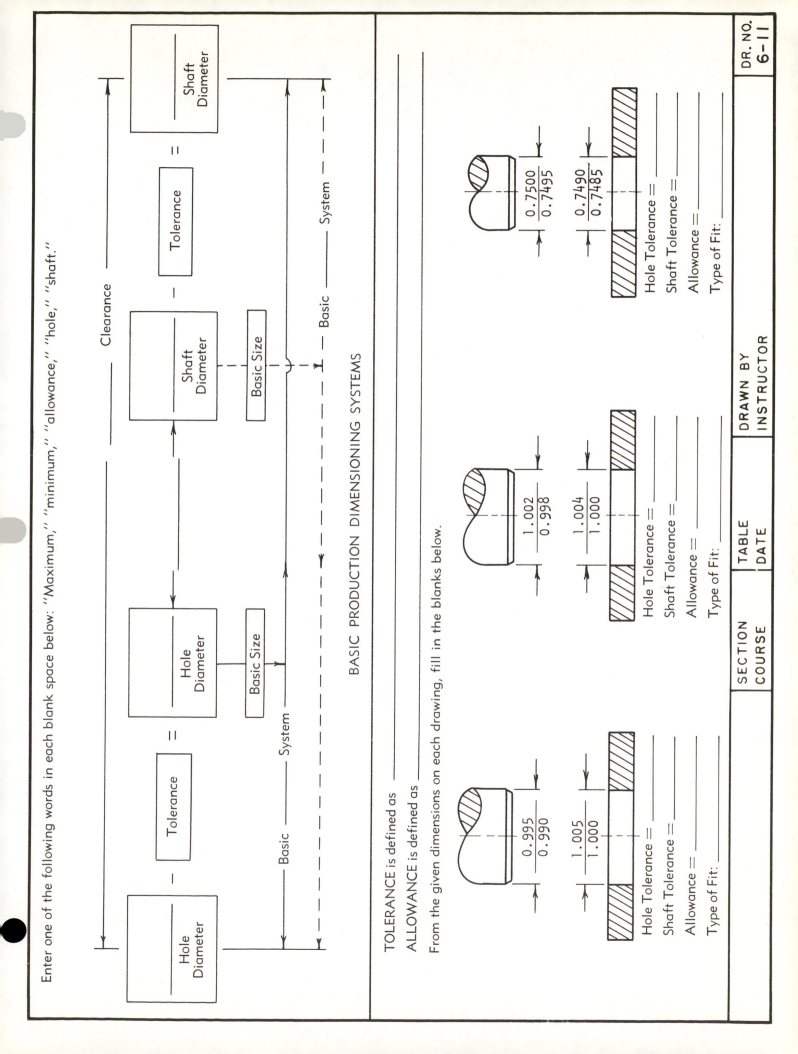

$\dfrac{0.995}{0.990}$ $\dfrac{1.005}{1.000}$

Hole Tolerance = _____
Shaft Tolerance = _____
Allowance = _____
Type of Fit: _____

$\dfrac{1.002}{0.998}$ $\dfrac{1.004}{1.000}$

Hole Tolerance = _____
Shaft Tolerance = _____
Allowance = _____
Type of Fit: _____

$\dfrac{0.7500}{0.7495}$ $\dfrac{0.7490}{0.7485}$

Hole Tolerance = _____
Shaft Tolerance = _____
Allowance = _____
Type of Fit: _____

SECTION	TABLE	DRAWN BY
COURSE	DATE	INSTRUCTOR

DR. NO.
6-11

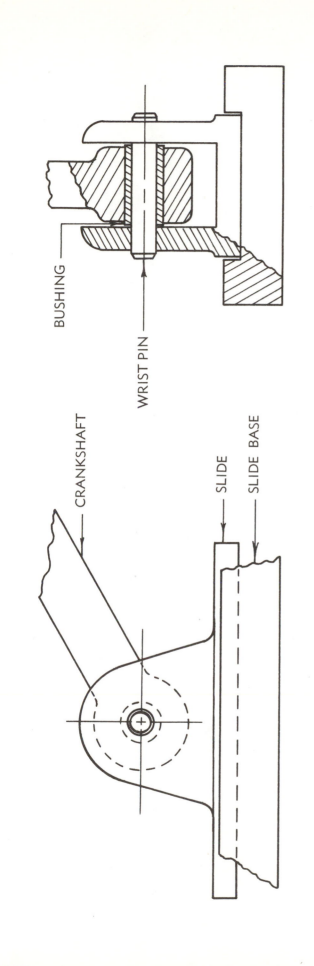

BUSHING

WRIST PIN

CRANKSHAFT

SLIDE

SLIDE BASE

PART	Nominal Size	Basic Size	Tolerance Zone	Hole Tolerance		Shaft Tolerance		Allowance
SLIDE BASE—Groove width	40		H9					
SLIDE—Tongue width	40		d9					
WRIST PIN—Diameter	6		h6					
WRIST PIN—Slide Hole Diam.	6		S7					
CRANKSHAFT—Hole Diameter	10		H7					
BUSHING—Outside diameter	10		n6					+0.010
BUSHING—Inside diameter	6							
WRIST PIN—Diameter								

SECTION	TABLE	DRAWN BY
COURSE	DATE	INSTRUCTOR

DR. NO.
6-12

Determine the limit dimensions indicated on the drawing of the adjustable pulley bracket. Use the basic shaft system for the shaft-bracket fit and the shaft-bushing fit. Use the basic hole system for the bushing-pulley fit and the bushing-bracket fit.

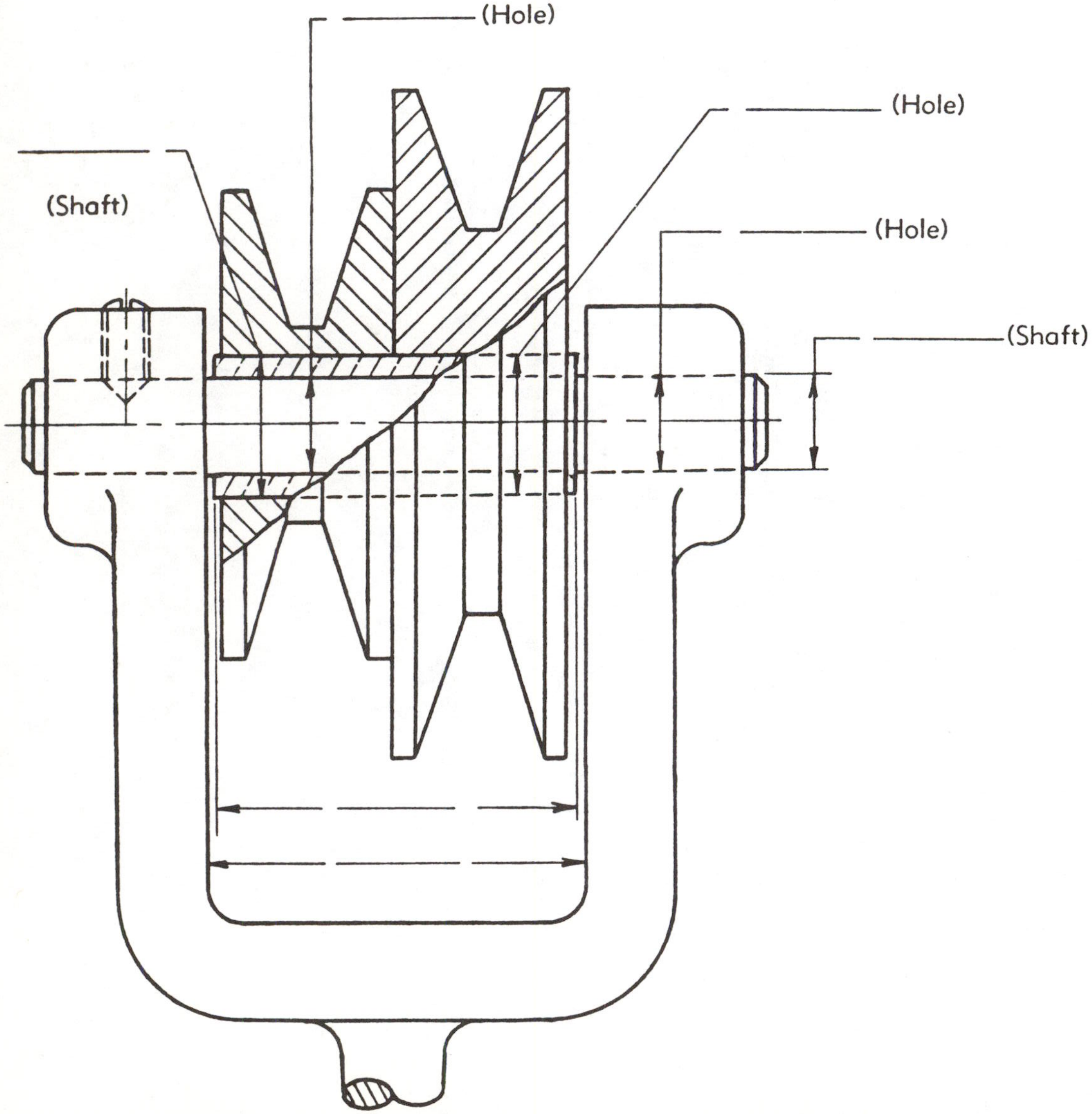

Specifications

SHAFT-BRACKET FIT

25 mm nominal diameter
0.052 shaft tolerance
0.052 hole tolerance
+0.065 allowance

SHAFT-BUSHING FIT

25 mm nominal diameter
0.052 shaft tolerance
0.021 hole tolerance
+0.007 allowance

BUSHING-PULLEY FIT

40 mm nominal diameter
0.025 hole tolerance
0.016 shaft tolerance
-0.059 allowance

BUSHING-BRACKET FIT

100 mm nominal length
0.220 hole tolerance
0.220 shaft tolerance
+0.170 minimum end play

Show computations below:

(Shaft)

(Hole)

(Hole)

(Shaft)

(Hole)

SECTION	TABLE	DRAWN BY
COURSE	DATE	INSTRUCTOR

In each problem, draw on the given center lines the type of thread indicated.

1. **Detail Thread Representation.** Do not show chamfer.

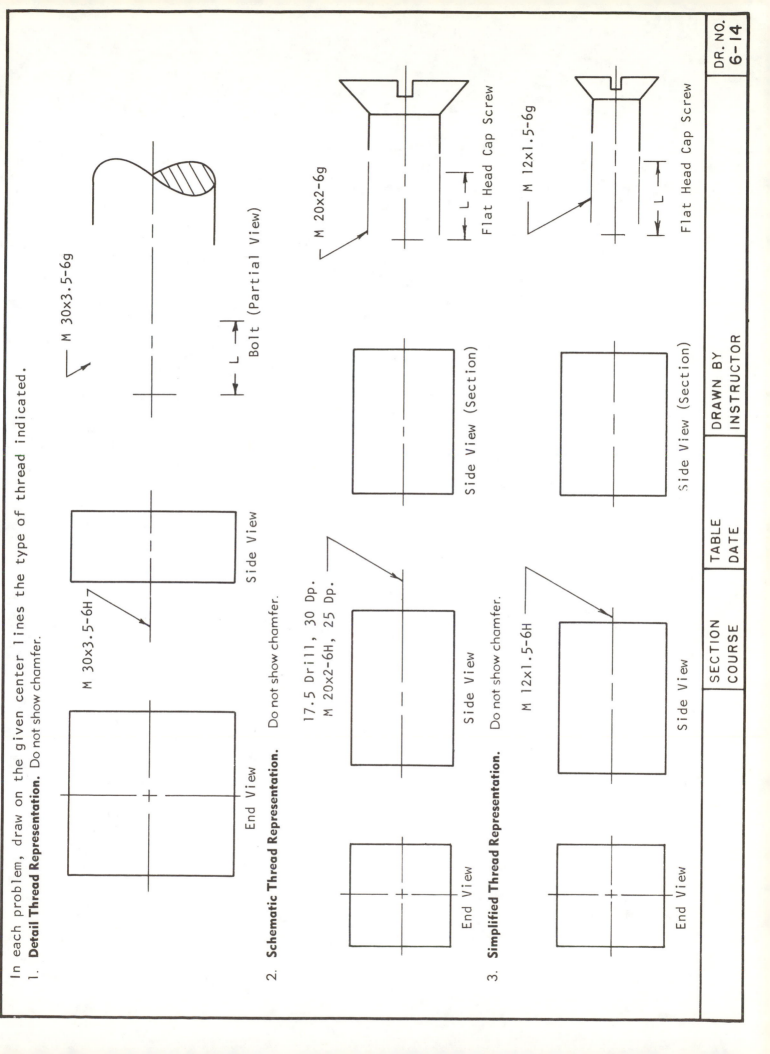

M 30x3.5-6g

Bolt (Partial View)

M 30x3.5-6H

End View   Side View

2. **Schematic Thread Representation.** Do not show chamfer.

M 20x2-6g

Flat Head Cap Screw

Side View (Section)

17.5 Drill, 30 Dp.
M 20x2-6H, 25 Dp.

End View   Side View

3. **Simplified Thread Representation.** Do not show chamfer.

M 12x1.5-6g

Flat Head Cap Screw

Side View (Section)

M 12x1.5-6H

End View   Side View

Draw a complete half size detail drawing of the cast iron bracket shown in the pictorial. Use a blank vellum sheet.

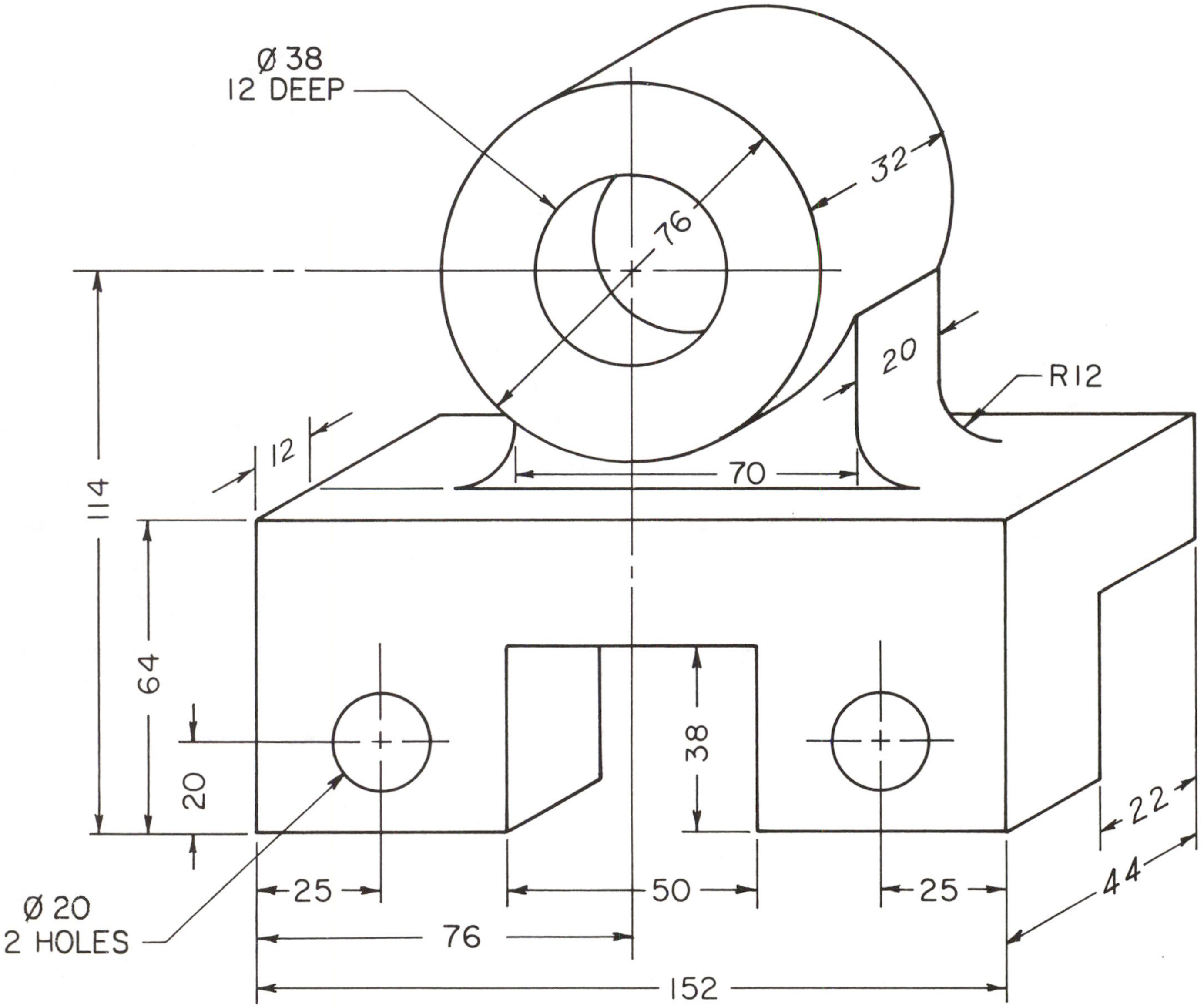

Ø38
12 DEEP

32

R12

20

70

76

Ø20
2 HOLES

12

20

64

114

38

50

76

25

152

25

44

22

| SECTION | TABLE | DRAWN BY | DR. NO. |
| COURSE | DATE | INSTRUCTOR | 7-1 |

Given the pictorial sketch of a Bearing Cap.
Make a detail drawing of the Cap.

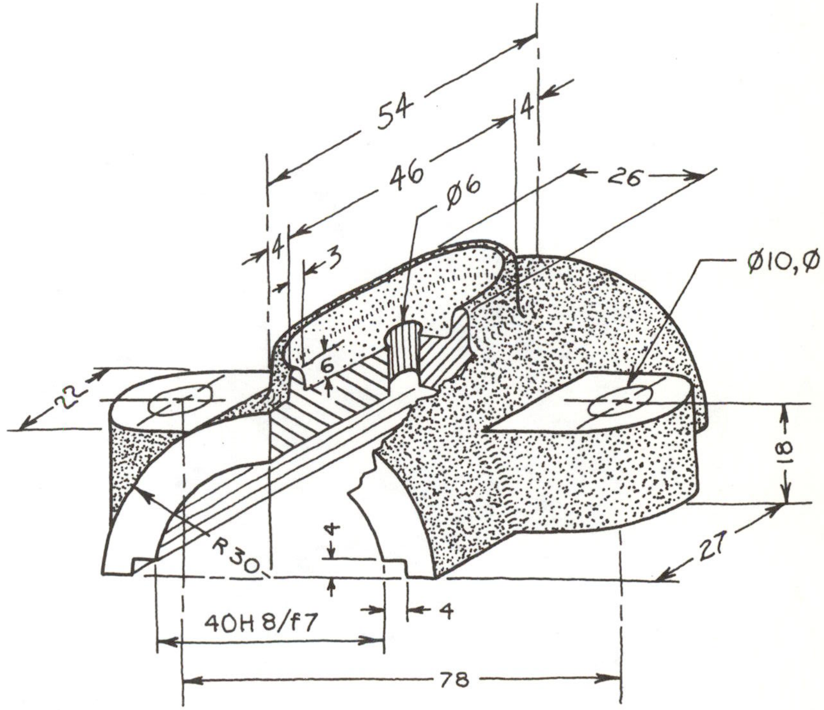

Ø10, Ø16 SF

81

27

26

54

46

Ø6

4

3

4

6

R30

4

4

78

40H 8/f7

22

Given the pictorial sketch of a Shaft Support Bracket.
Make a detail drawing of the bracket.

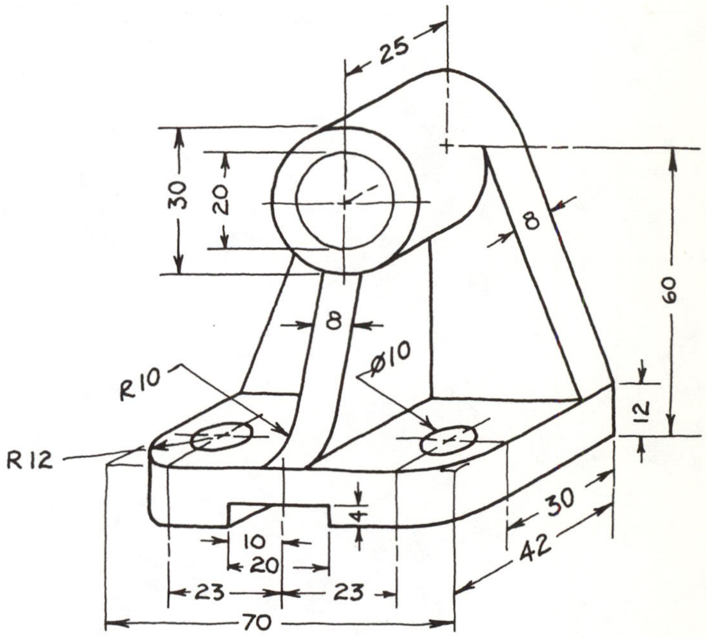

SECTION	TABLE	DRAWN BY
COURSE	DATE	INSTRUCTOR

Given the dimensioned pictorial view of a Shaft End Support Bracket.
Make a detail drawing of the bracket. Scale: Half size.

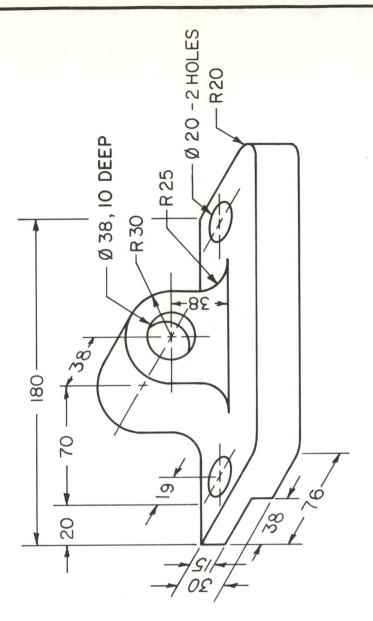

SECTION	TABLE	DRAWN BY
COURSE	DATE	INSTRUCTOR

The design drawing showing both details and a pictorial assembly of a machinist's bench vise is reproduced here, courtesy of the Casting Specialties Corporation, Cedarburg, Wisconsin, 53012. The designer usually gives only the critical dimensions on a design drawing. The person who draws detail drawings of the various parts and the necessary assembly drawing is expected to use good proportions, scaling if necessary the design drawing. As the detailer, make detail drawings of the parts or an assembly drawing as assigned. Change the specifications to SI for international as well as domestic manufacture. Use preferred sizes where appropriate, otherwise round off to the nearest even number of millimeters.

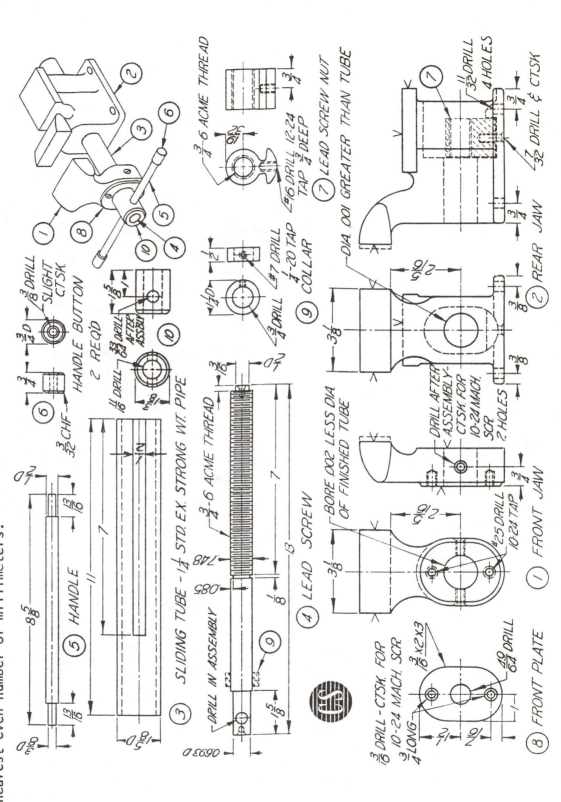

| SECTION | TABLE | DRAWN BY |
| COURSE | DATE | INSTRUCTOR |

The design drawing showing both details and a pictorial assembly of a machinist's bench vise is reproduced here, courtesy of the Casting Specialties Corporation, Cedarburg, Wisconsin, 53012. The designer usually gives only the critical dimensions on a design drawing. The person who draws detail drawings of the various parts and the necessary assembly drawing is expected to use good proportions, scaling if necessary the design drawing. As the detailer, make detail drawings of the parts or an assembly drawing as assigned. Change the specifications to SI for international as well as domestic manufacture. Use preferred sizes where appropriate, otherwise round off to the nearest even number of millimeters.

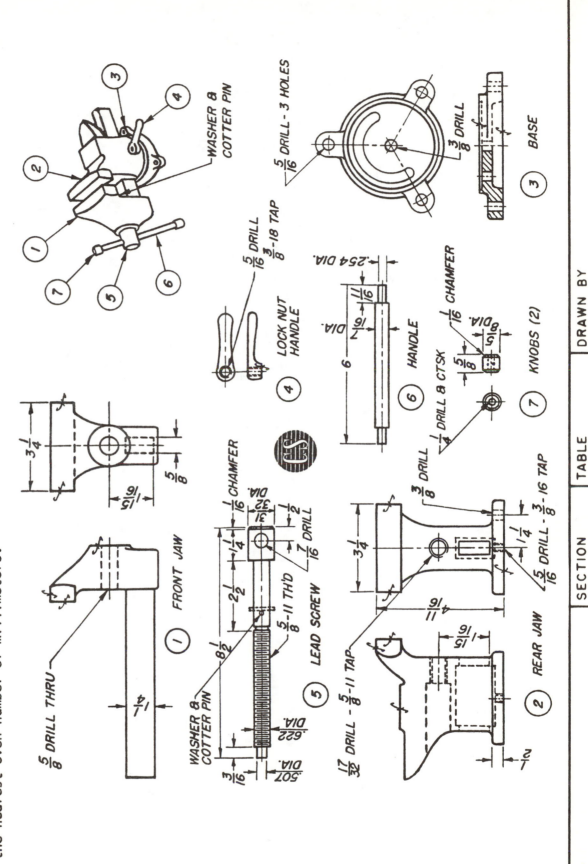

SECTION	TABLE	DRAWN BY
COURSE	DATE	INSTRUCTOR

Problem 1. Determine the resultant of the four forces acting on the plate. Vector Scale: 1 cm = 5 Newtons

Problem 2-6. Determine the magnitude and nature of stress in each member of the given frames. Use Bow's notation and sketch free-body diagrams. Indicate tension by T and compression by C.
Vector Scale: 1 cm = 5 Newtons

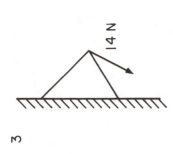

1

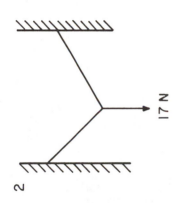

2

17 N

3

14 N

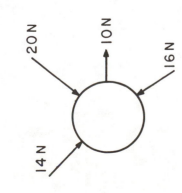

4

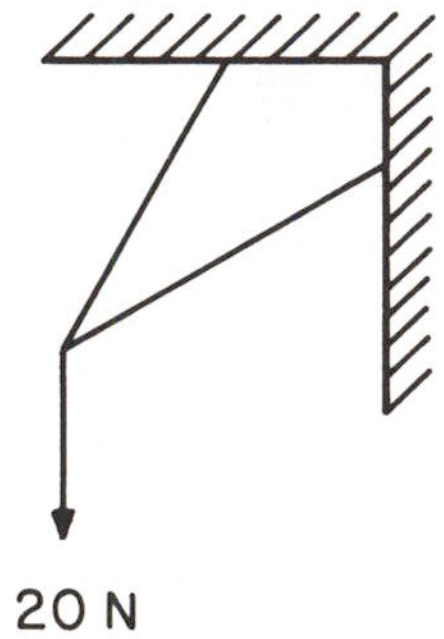

20 N

5

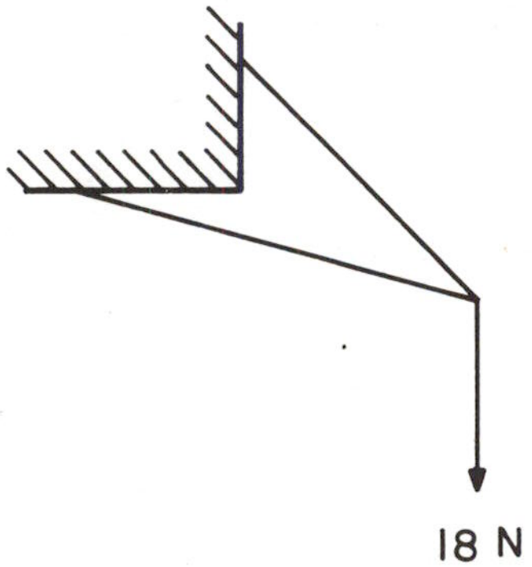

18 N

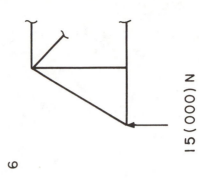

6

15(000) N

SECTION

COURSE

TABLE

DATE

DRAWN BY
INSTRUCTOR

DR. NO.
8-1

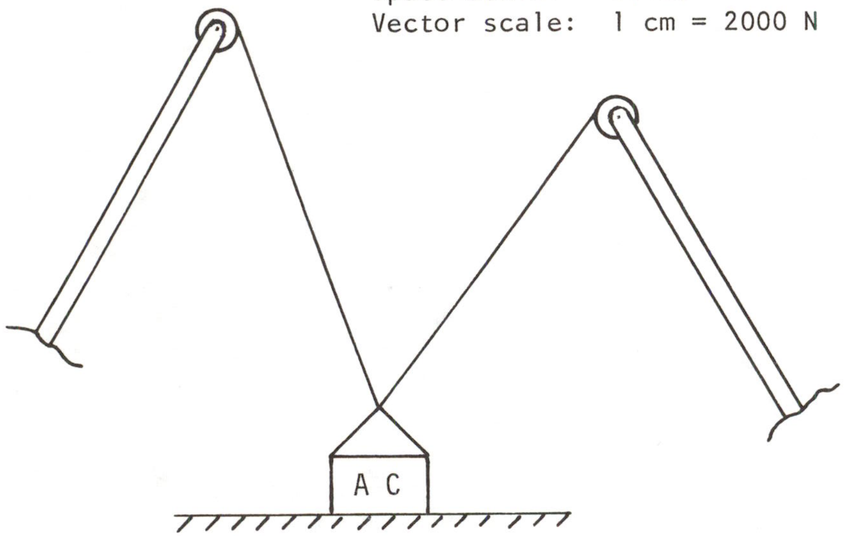

A typical light aircraft is loaded as shown during
engine run-up on the ground. Determine the reactions
on the nose wheel and on each main wheel. Brakes are
applied on the main wheels only.

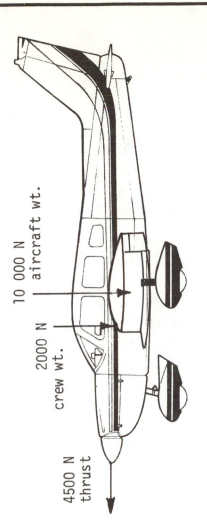

10 000 N
aircraft wt.

2000 N
crew wt.

4500 N
thrust

How high can a 10 000 N air conditioning package be
raised if the crane cable on the left has a load
limit of 9000 N and the crane cable on the right
has a load limit of 8000 N?

Space scale: 1:100
Vector scale: 1 cm = 2000 N

A C

DR. NO.
8-2

DRAWN BY
INSTRUCTOR

TABLE
DATE

SECTION
COURSE

Given the truss loaded as shown. Determine graphically the reactions at the supports and the magnitude and nature of the stress in each member for joints 1, 2 and 3. Apply Bow's notation. Indicate the nature of the stress with the letters T for tension and C for compression and tabulate the results.

Vector Scale: 1 cm = 200 Newtons

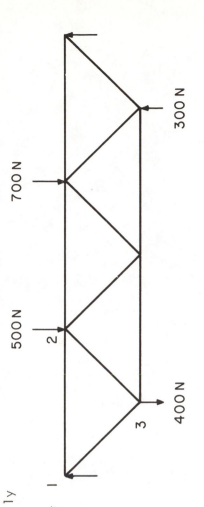

700 N

500 N

2

300 N

400 N

3

SECTION	TABLE	DRAWN BY	DR. NO.
COURSE	DATE	INSTRUCTOR	8-3

Given the truss loaded as indicated. Determine graphically the magnitude and nature of stress in each member and also the magnitude and line of action of the reactions at the roller and the pinned joint. Use Bow's notation as given. Enter the results in the table below and draw the lines of action of the reactions on the drawing of the truss.

Vector Scale: 1 cm = 1000 N

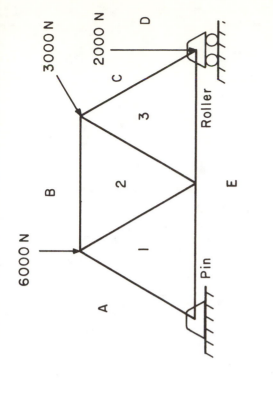

Force	Magnitude	Nature of Stress
A1	____	____
B2	____	____
C3	____	____
E1	____	____
E3	____	____
12	____	____
23	____	____
DE	____	____
EA	____	____

SECTION	TABLE	DRAWN BY	DR. NO.
COURSE	DATE	INSTRUCTOR	8-4

For each of the force systems below, determine the magnitude and direction of the equilibrant (E).

A = 30 N
B = 35 N
C = 40 N
Vector Scale: 1 cm = 10 N

A = 160 N
B = 70 N
C = 60 N
D = 180 N
Vector Scale: 1 cm = 40 N

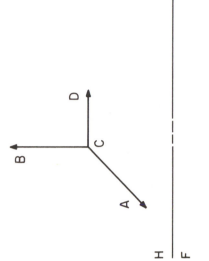

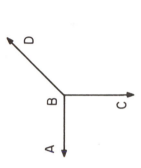

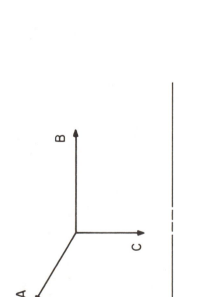

E = _____

Bearing = _____

Slope Angle = _____

E = _____

Bearing = _____

Slope Angle = _____

SECTION	TABLE	DRAWN BY
COURSE	DATE	INSTRUCTOR

Determine the forces in each of the three cables supporting the 600 Newton load.   1 cm = 100 N

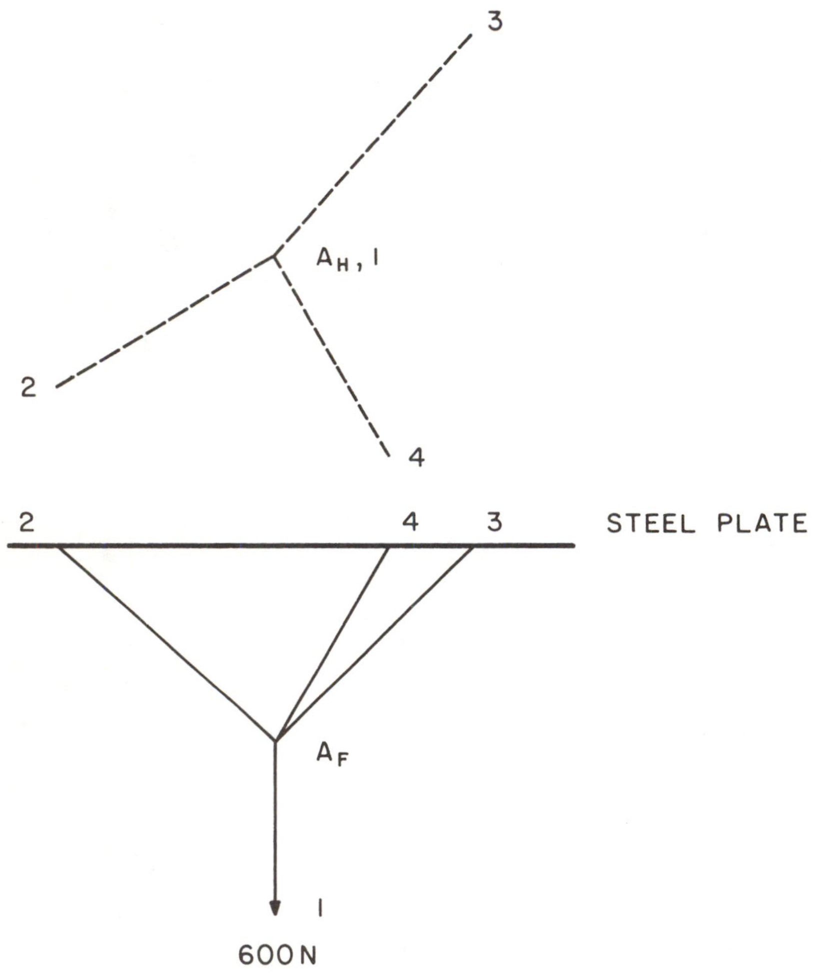

3

$A_H, 1$

2

4

STEEL PLATE

3

4

A_F

1

2

600 N

600

| SECTION | TABLE | DRAWN BY |
| COURSE | DATE | INSTRUCTOR |

DR. NO.
8 - 6

Design a multiple bar chart to graphically display the data collected to support an airport expansion proposal. Use a sheet of plain 8 1/2" x 11" vellum paper.

Business Jets and Heavy Twin Engine Aircraft (FAA Type C)

Based Aircraft, Number of Operations, Fuel Consumption, Inspection and Maintenance Expenditures

USA County Airport

	1975	1980	1985
Number of Aircraft			
Total Type C Aircraft	10	18	29
Percent of Total Based Aircraft	2.4%	3.0%	3.4%
Number of Operations			
Total Type C Operations	6,600	12,800	20,900
Percent of Total Operations	3.5%	4.2%	4.4%
Fuel Consumption gallons (x1000)			
Type C Aircraft	1,267	2,471	4,180
Percent of Total Fuel Consumption	58%	63%	66%
Inspection and Maintenance Expenditures			
Type C Aircraft	$245,000	$458,000	$749,000
Percent of Total I&M Expenditures	39%	41%	43%

SECTION	TABLE	DRAWN BY	DR. NO.
COURSE	DATE	INSTRUCTOR	9-1

Plot the following data and draw the graph of the curve. Graphically determine the equation of the curve using an appropriate logarithmic grid.

X	0.0	1.0	2.0	3.0	4.0	5.0	6.0	7.0	8.0
Y	0.0	1.7	9.0	26.5	54.4	95.0	149.9	220.4	307.7

Plot the following data and draw the graph of
the curve, and determine the equation of the
curve using an appropriate logarithmic grid.

X	1.0	2.3	3.1	3.9	4.4	5.2	6.0	7.4
Y	4.4	2.3	1.5	10.	0.8	0.5	0.4	0.2

SECTION	TABLE	DRAWN BY
COURSE	DATE	INSTRUCTOR

Time vs. distance data for a particle moving on a straight line
with constant acceleration is as follows:

Time (s)	1.5	2.0	3.5	5.0	6.0	7.5	9.0	10.0
Distance (m)	1.35	2.40	7.35	15.00	21.60	33.75	48.60	60.00

Plot a graph of the data and determine the equation of the curve.
What is the acceleration of the particle?

DRAWN BY
INSTRUCTOR

TABLE
DATE

SECTION
COURSE

A 45%/55% refrigerant/oil mixture is tested for viscosity at various temperatures and the results tabulated as shown. Viscosity is measured in Saybolt Seconds Universal, the time required for a standard volume of the oil to flow through a standard orifice. Plot the data and determine the equation for the curve.

Temp. (°F)	-40	-20	0	20	40	60	80	100	120
Viscosity (SSU)	90	76	66	56	49	42	36	32	27

DR. NO. 9-5

DRAWN BY

INSTRUCTOR

TABLE

DATE

SECTION

COURSE

A particle starts 12 m to the
left of an origin (-12 m) and
moves along a straight line
path in accordance with the
v-t curve shown at the right.

Plot the s-t and a-t curves.

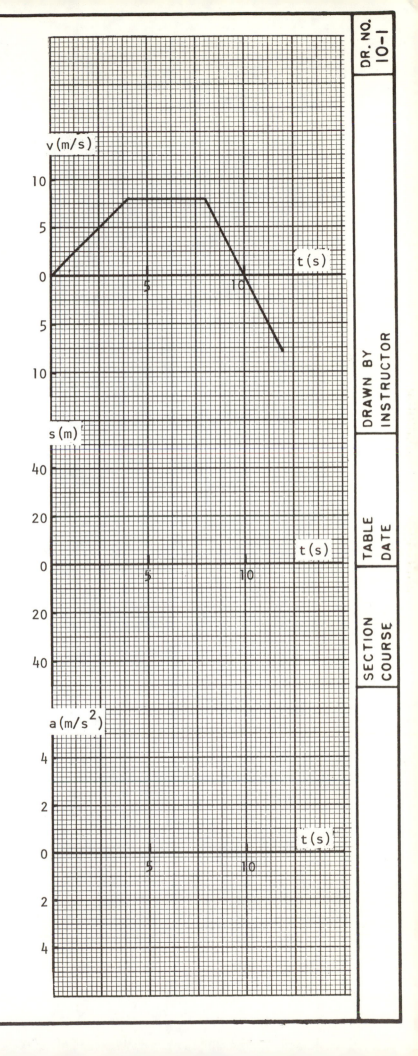

Given the s-t curve below, plot the v-t curve and determine the equations for the curve segments.

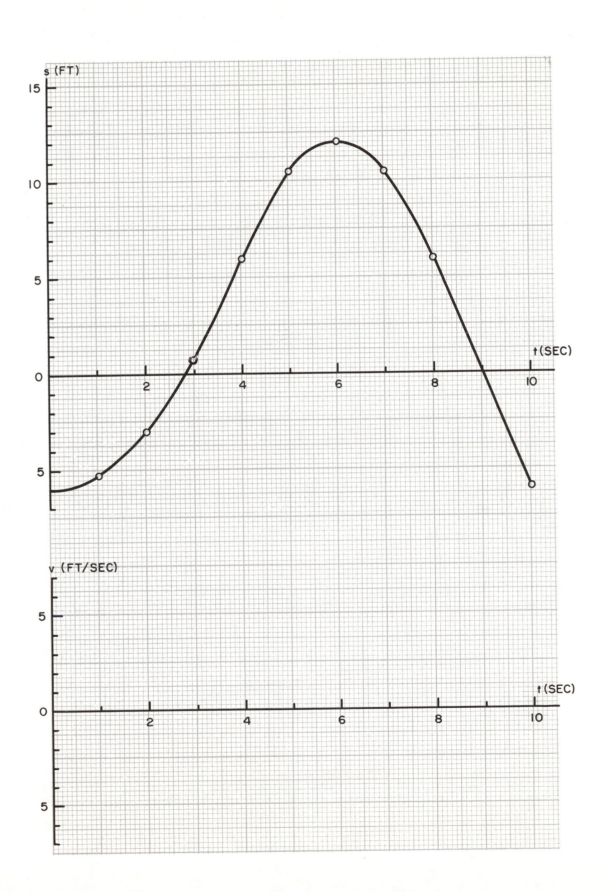

A plot of land, drawn to scale below, lies between a stream and three streets. Using graphical calculus, determine the area of the plot (m²) and divide the plot into four equal areas with North-South interior boundary lines.

Scale 1:1000

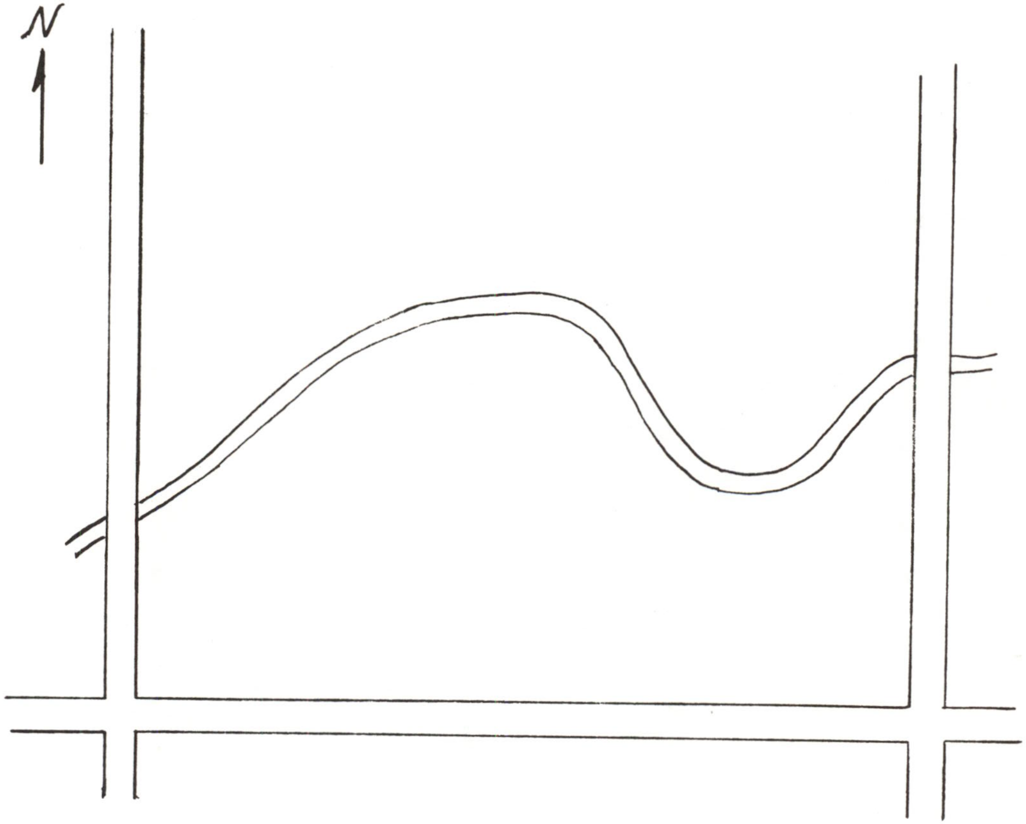

SECTION
COURSE

TABLE
DATE

DRAWN BY
INSTRUCTOR

DR. NO.
10-3

Plot the equation $Y = X^4 - 8X^3 + 10X^2 + 100X - 50$ from $X = 1$ to $X = 7$.

Determine the area under the curve using (1) the trapezoidal rule, (2) Simpson's rule, and (3) the calculus.

SECTION	TABLE
COURSE	DATE

DRAWN BY
INSTRUCTOR

DR. NO.
10-4

DRAWN BY
INSTRUCTOR

TABLE
DATE

SECTION
COURSE

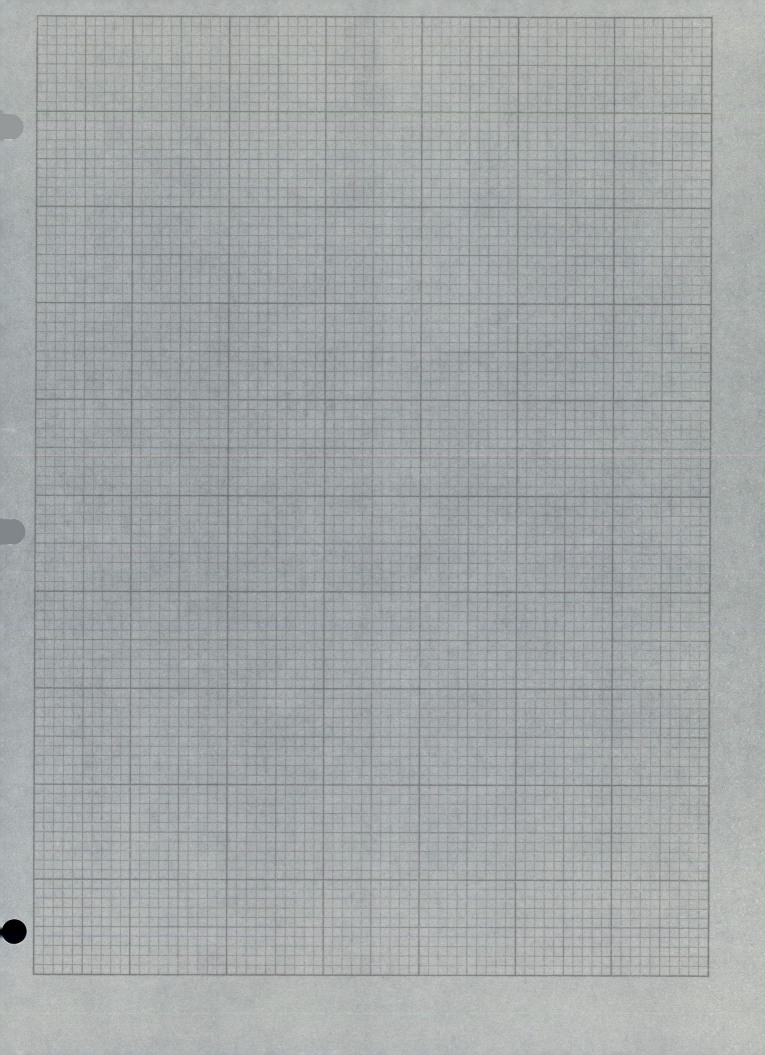

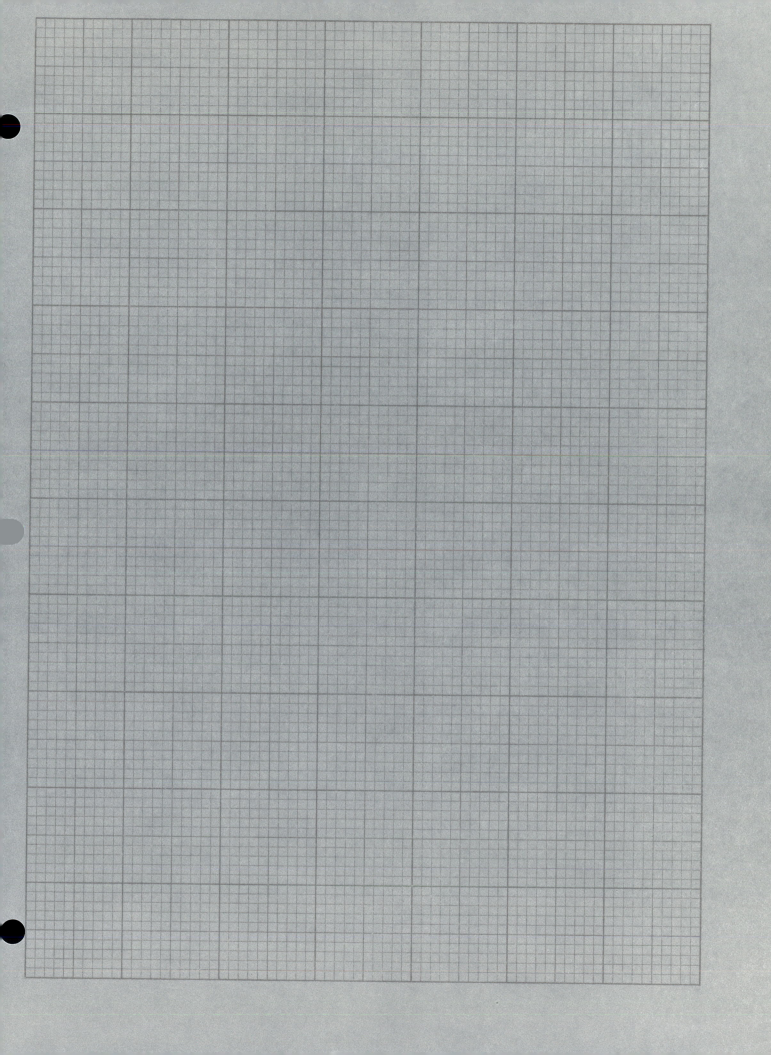

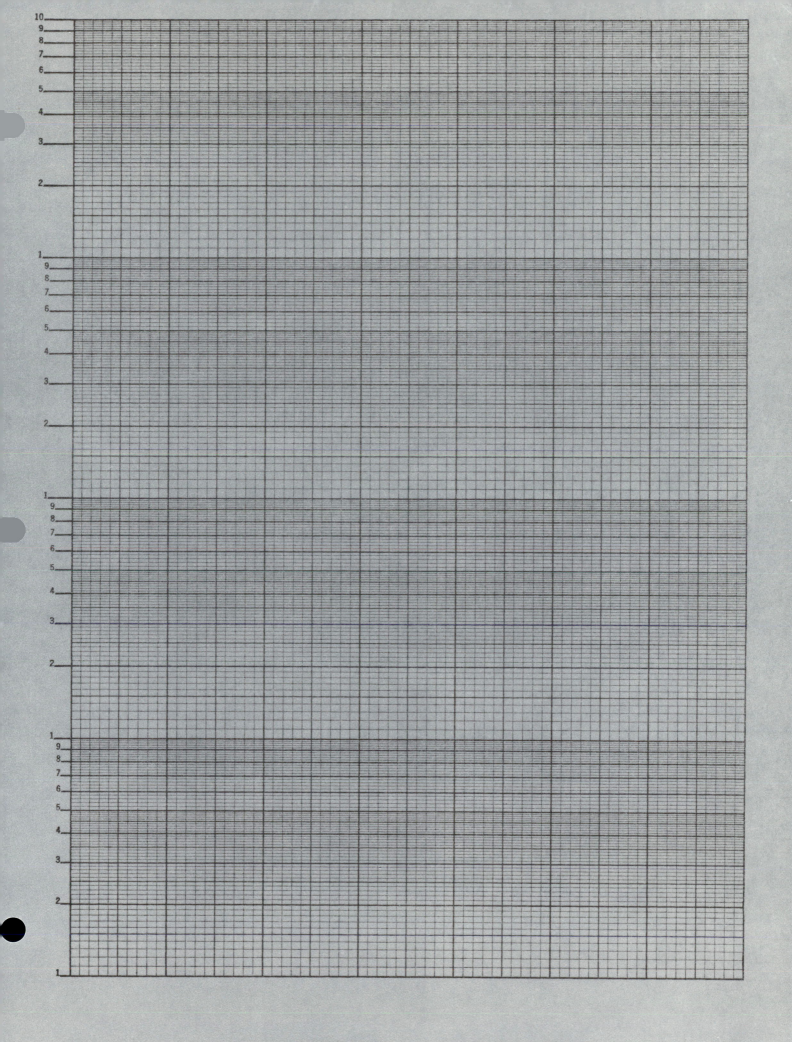

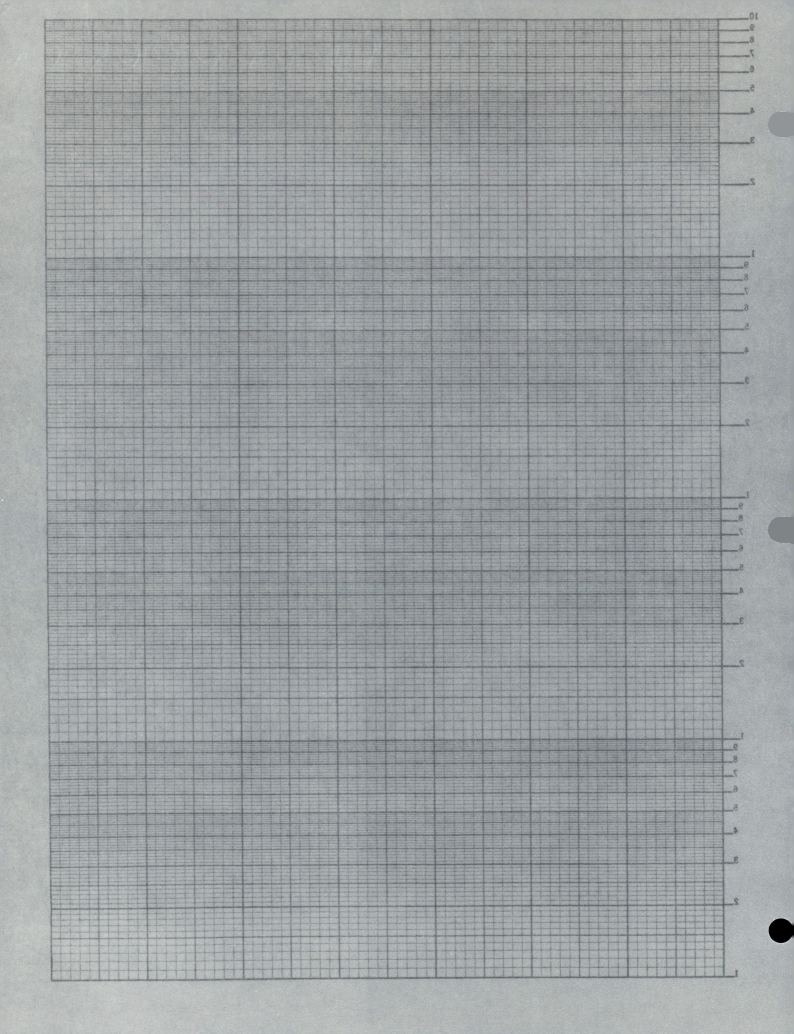

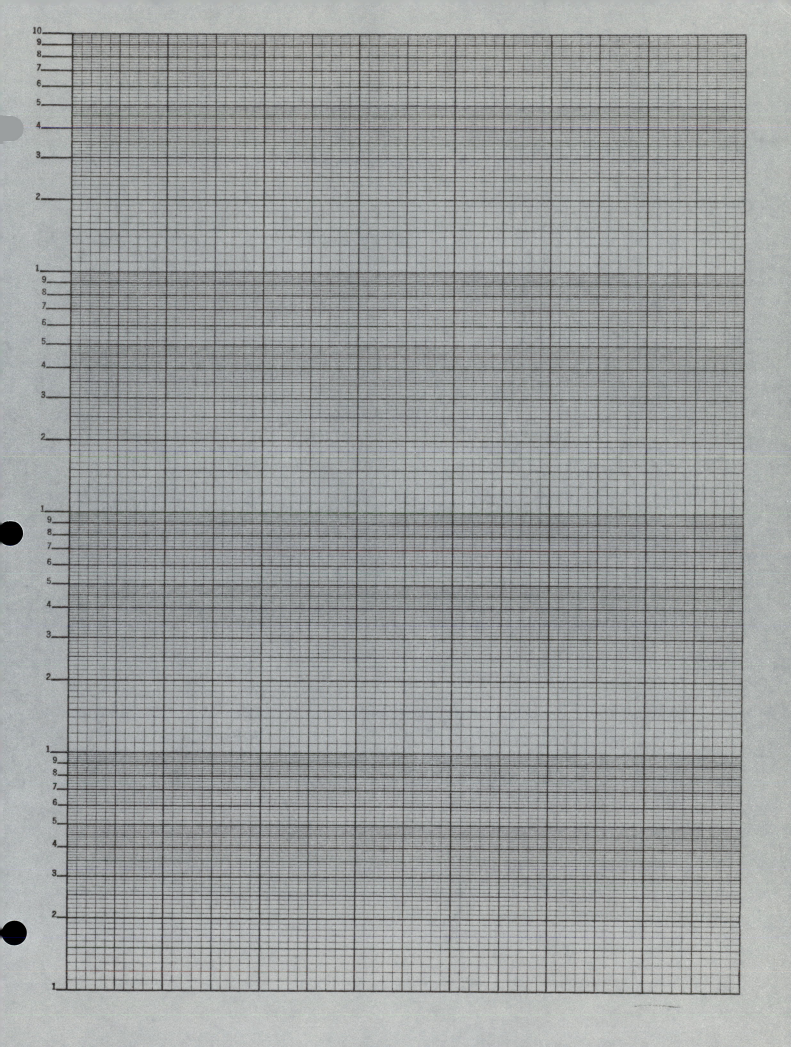

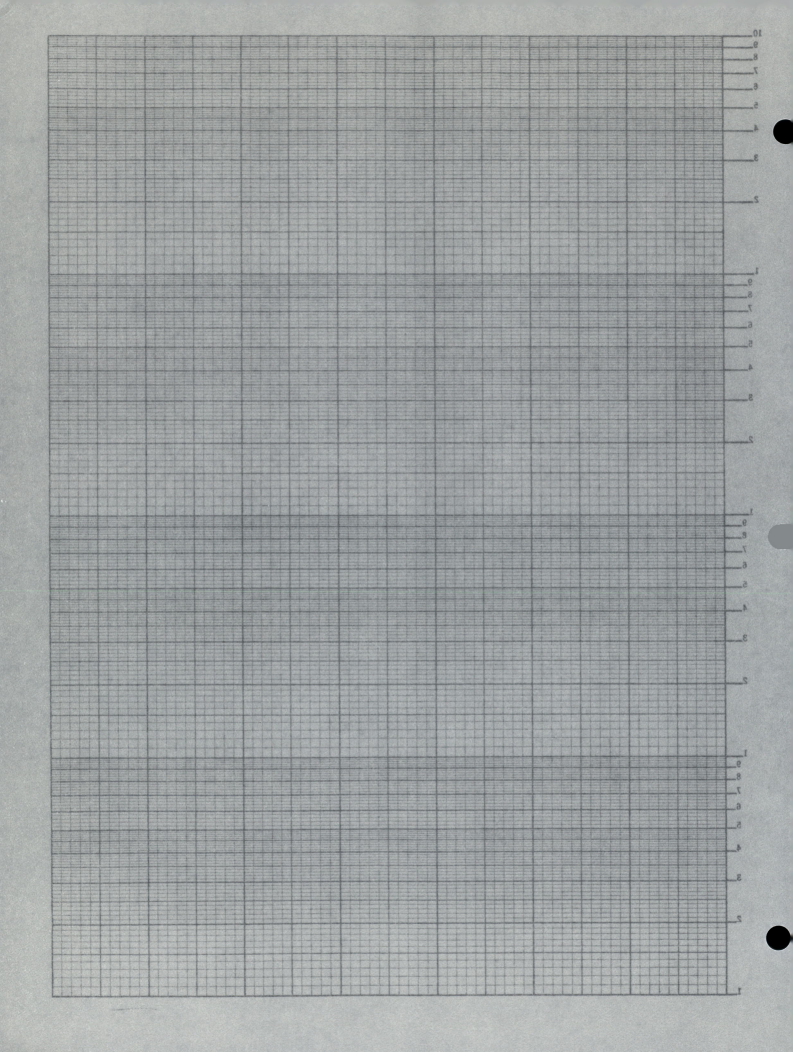

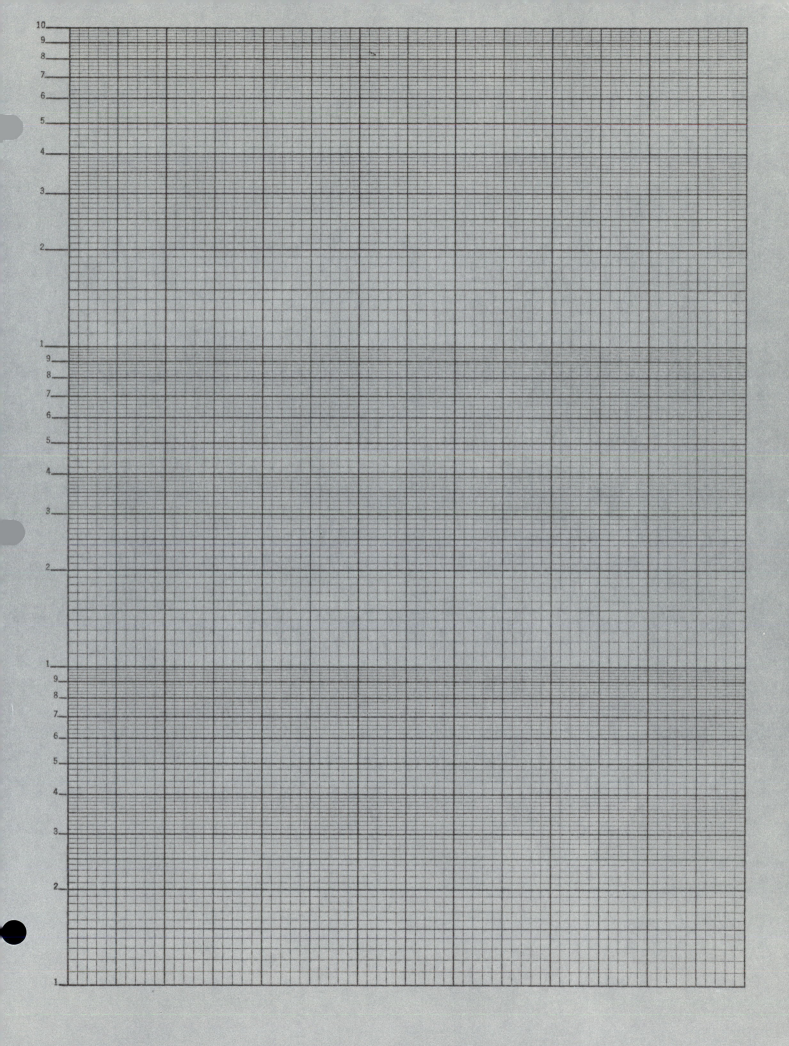

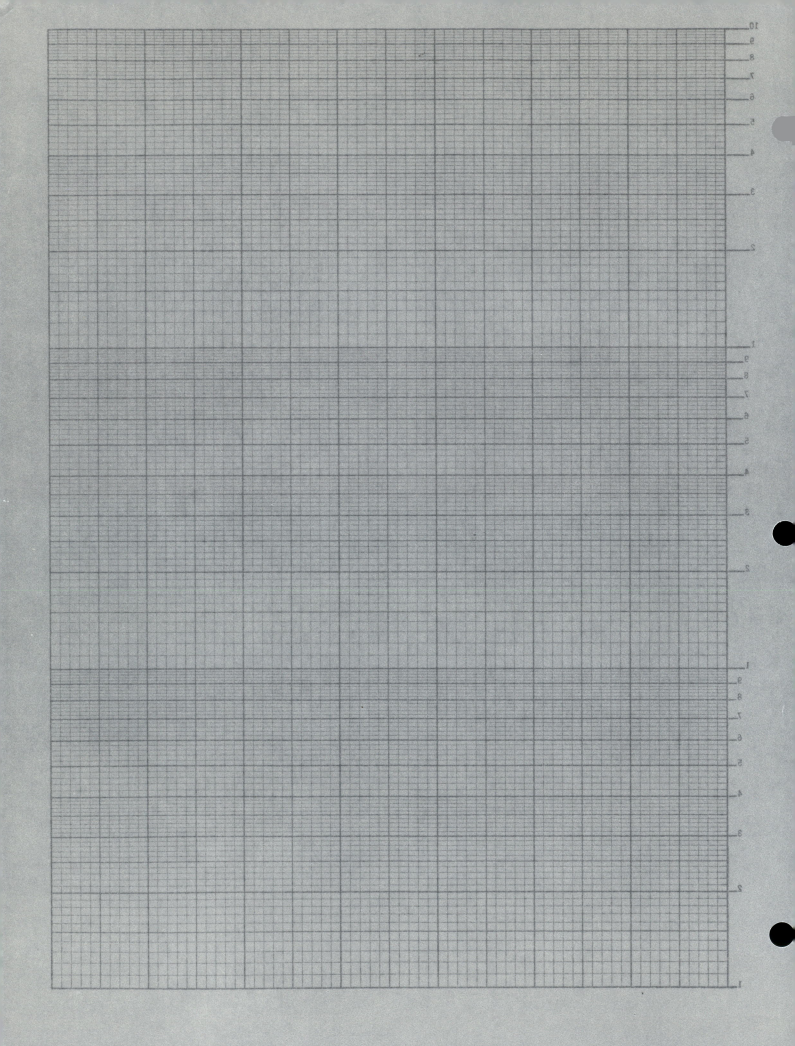

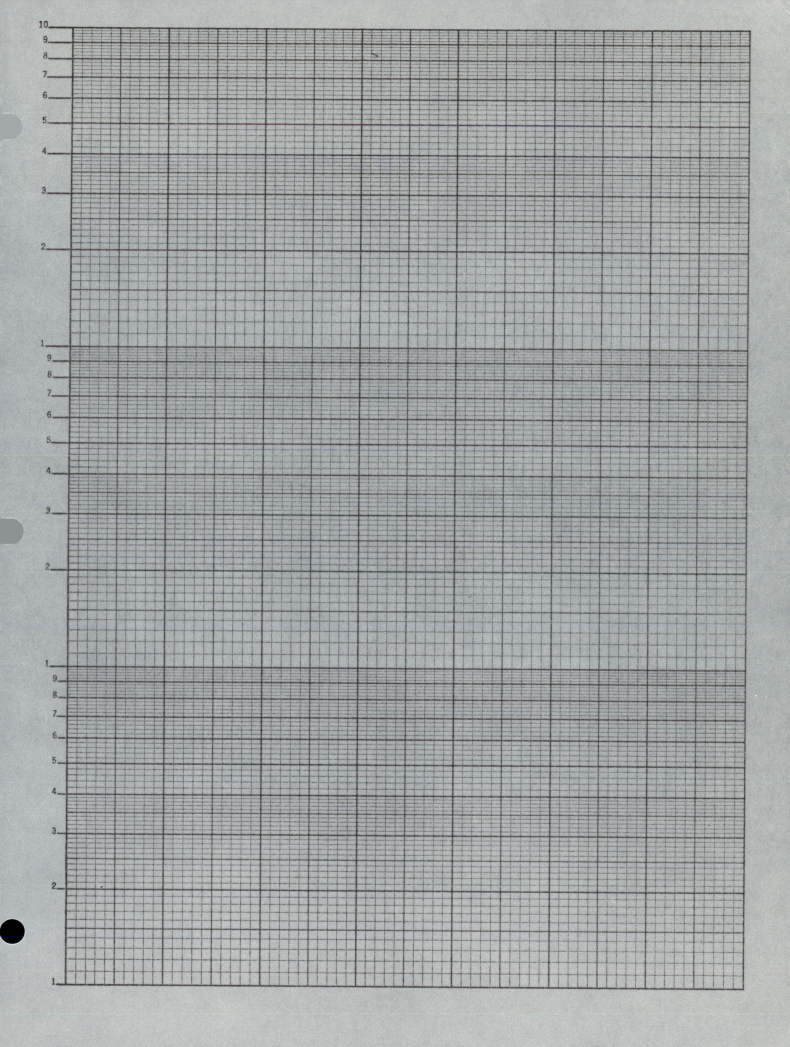

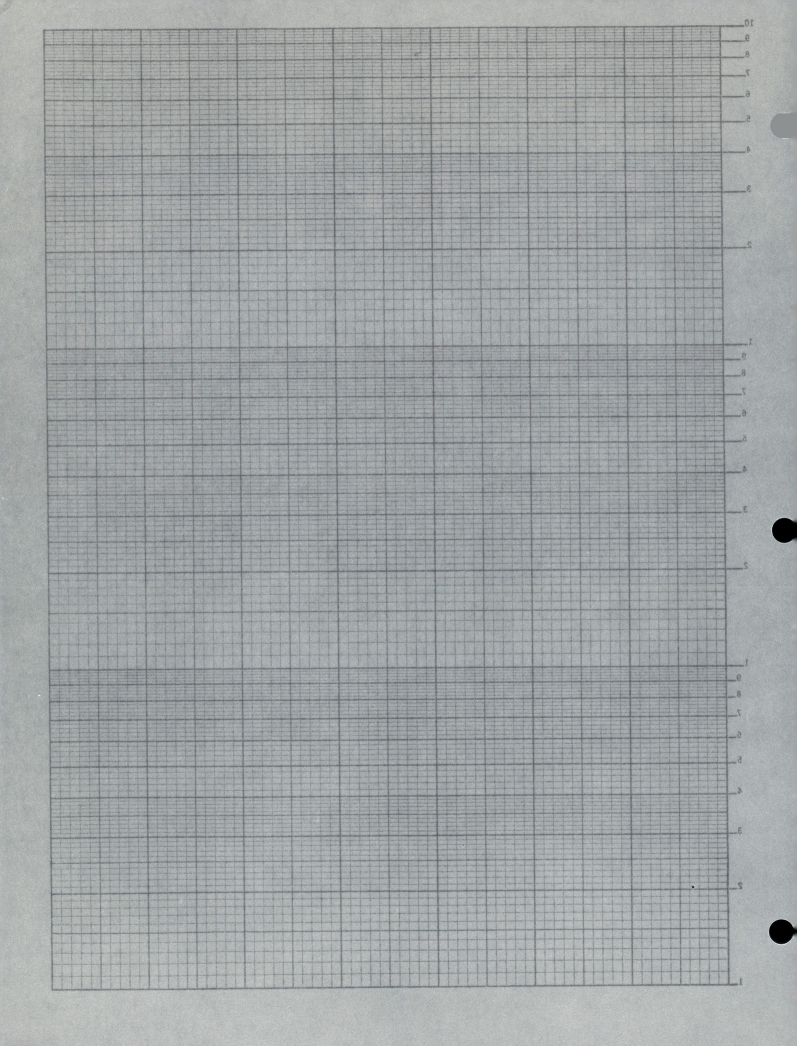

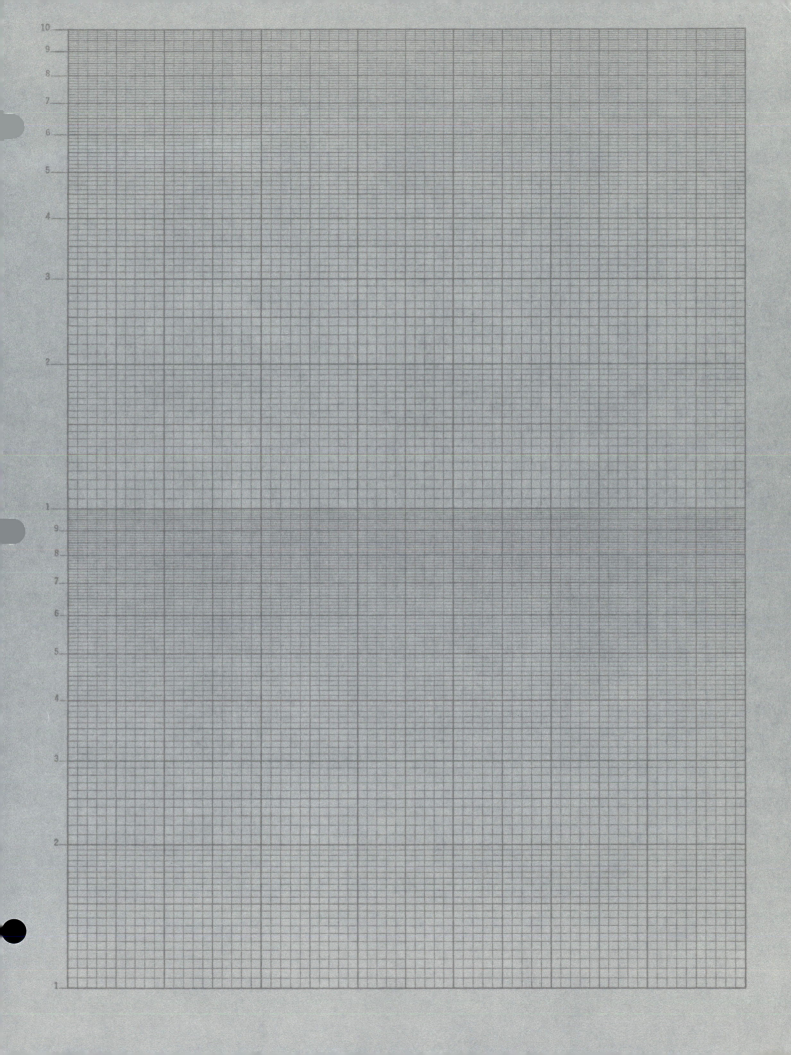

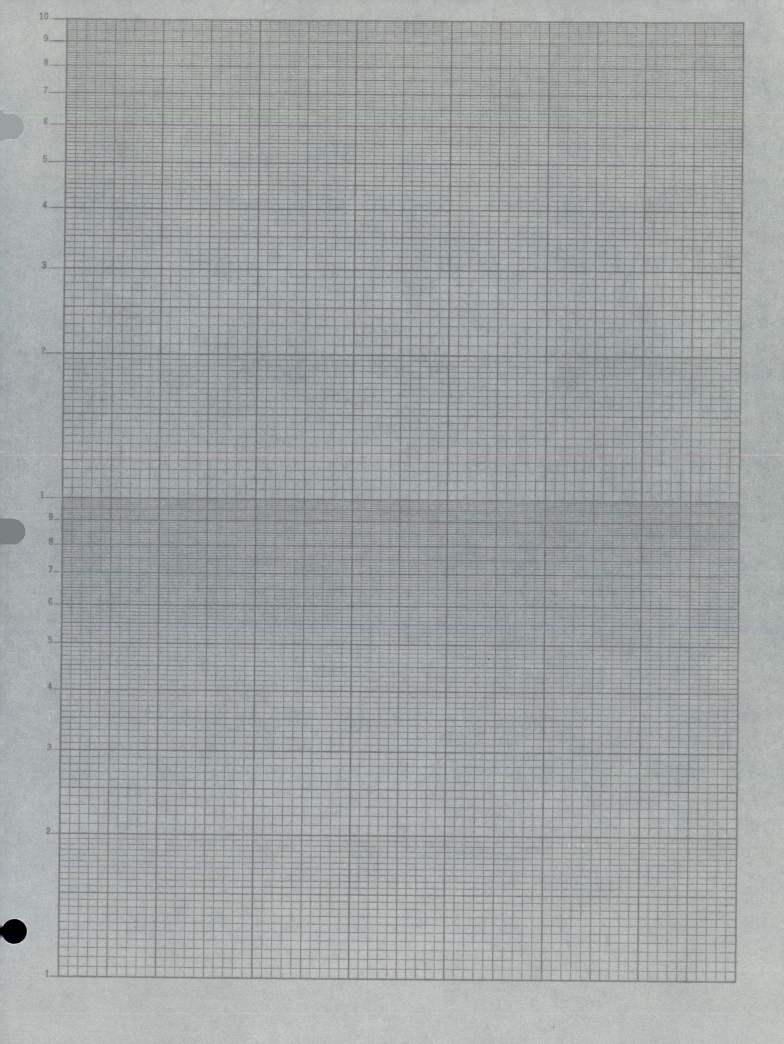

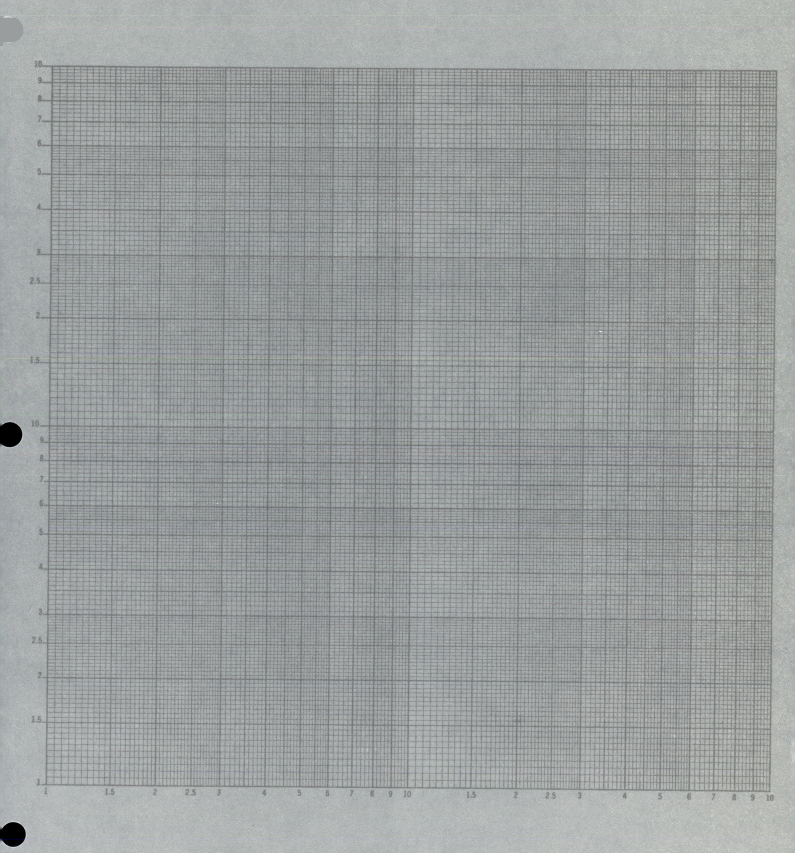

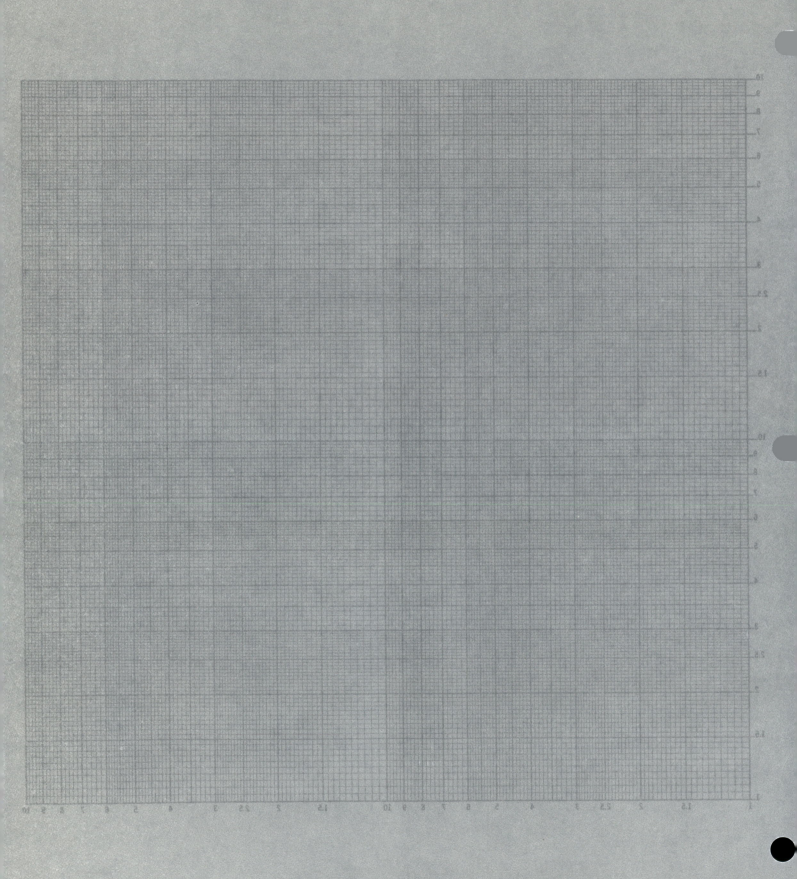

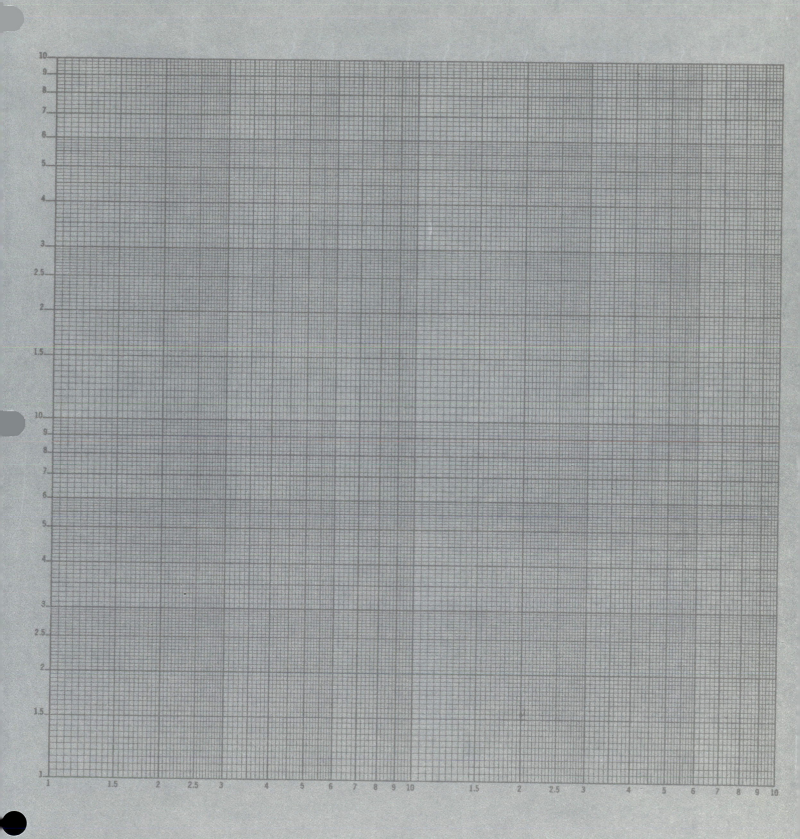

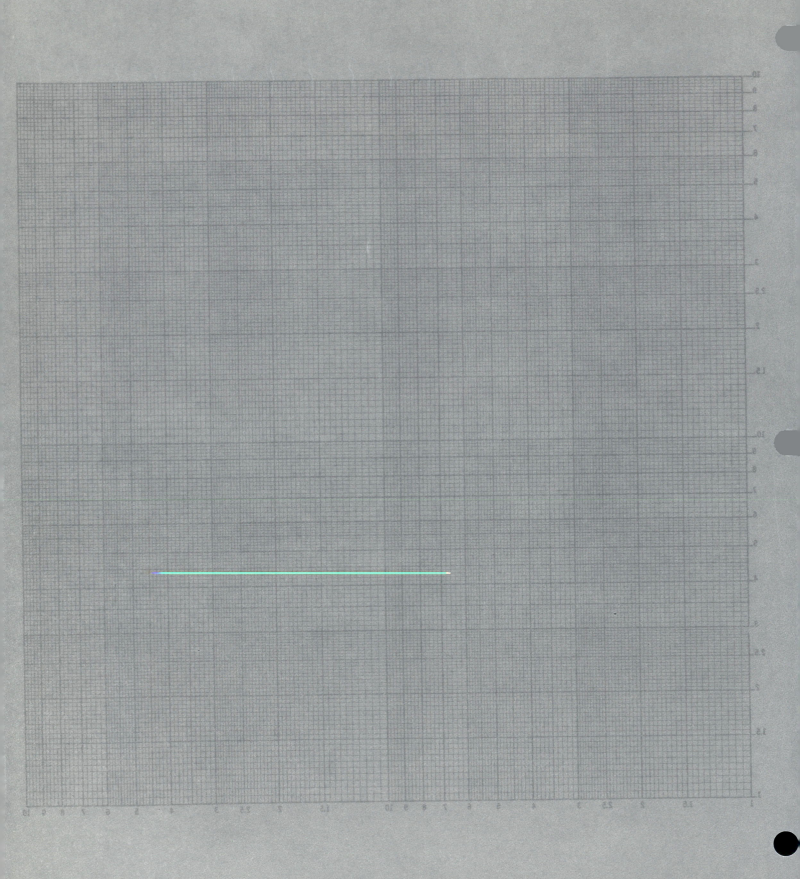

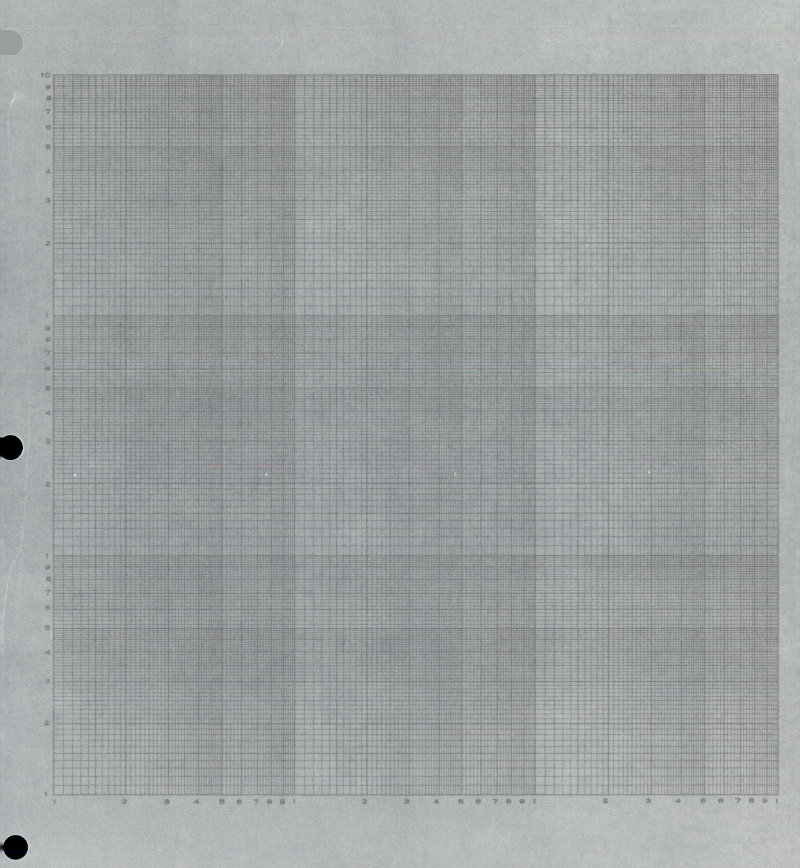

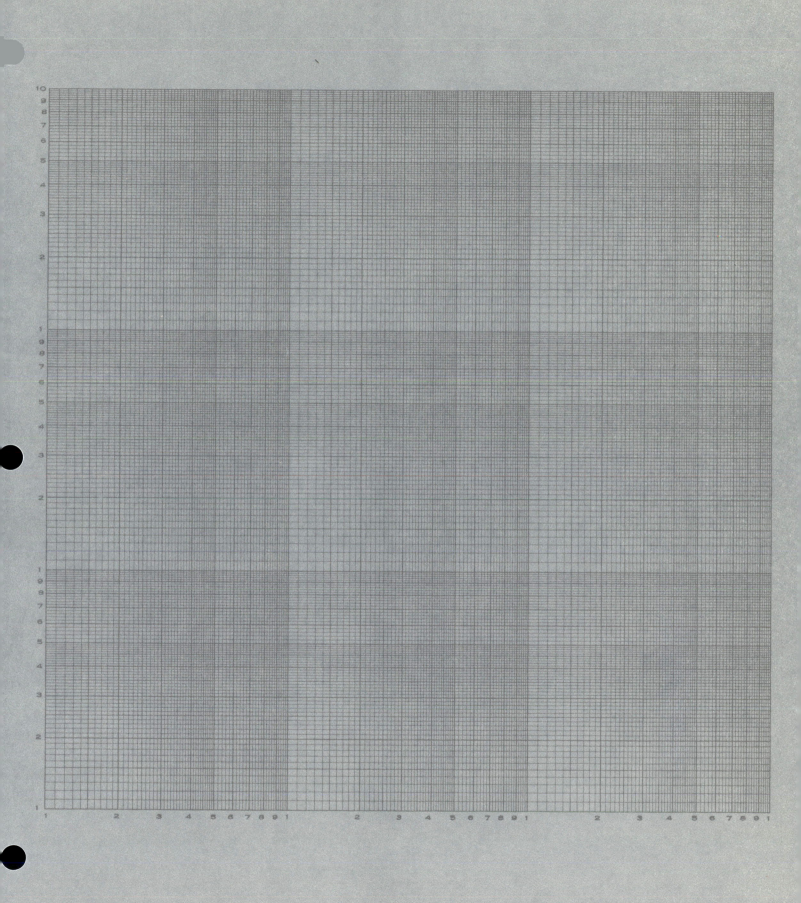